T0337829

The Practice of Engineering Dynamics

The Practice of Engineering Dynamics

Ronald J. Anderson
Queen's University
Kingston, Canada

This edition first published 2020
© 2020 John Wiley & Sons Ltd

The right of Ronald J. Anderson to be identified as the author of this work has been asserted in accordance with law.

Registered Offices
John Wiley & Sons, Inc., 111 River Street, Hoboken, NJ 07030, USA
John Wiley & Sons Ltd, The Atrium, Southern Gate, Chichester, West Sussex, PO19 8SQ, UK

Editorial Office
The Atrium, Southern Gate, Chichester, West Sussex, PO19 8SQ, UK

For details of our global editorial offices, customer services, and more information about Wiley products visit us at www
.wiley.com.

Wiley also publishes its books in a variety of electronic formats and by print-on-demand. Some content that appears in
standard print versions of this book may not be available in other formats.

Library of Congress Cataloging-in-Publication data

Names: Anderson, Ron J. (Ron James), 1950- author.
Title: The practice of engineering dynamics / Ronald J Anderson, Queen's
    University, Kingston, Canada.
Description: First edition. | Hoboken, NJ, USA : John Wiley & Sons, Inc.,
    [2020] | Includes bibliographical references and index.
Identifiers: LCCN 2020004354 (print) | LCCN 2020004355 (ebook) | ISBN
    9781119053705 (hardback) | ISBN 9781119053682 (adobe pdf) | ISBN
    9781119053699 (epub)
Subjects: LCSH: Machinery, Dynamics of. | Mechanics, Applied.
Classification: LCC TJ170 .A53 2020 (print) | LCC TJ170 (ebook) | DDC
    620.1/04–dc23
LC record available at https://lccn.loc.gov/2020004354
LC ebook record available at https://lccn.loc.gov/2020004355

Cover Design: Wiley
Cover Image: © kovop58/Shutterstock

Set in 9.5/12.5pt STIXTwoText by SPi Global, Chennai, India

Printed and bound by CPI Group (UK) Ltd, Croydon, CR0 4YY

10 9 8 7 6 5 4 3 2 1

To June, Stacey, and Kate

Contents

Preface

The design of a mechanical system very often includes a requirement for dynamic analysis. During the early concept design stages it is useful to create a mathematical model of the system by deriving the governing equations of motion. Then, simulations of the behavior of the system can be produced by solving the equations of motion. These simulations give guidance to the design engineers in choosing parameter values in their attempt to create a system that satisfies all of the performance criteria they have laid out for it.

There is a logical progression of analyses that are required during the design. The design engineer needs to determine, from the nonlinear differential equations of motion:

- The equilibrium states of the system – these are places where, once put there and not disturbed, the system will stay. The time varying terms are removed from the differential equations of motion, leaving a set of nonlinear algebraic equations. The solutions to these equations provide knowledge of all of the equilibrium states.

- The stability of the equilibrium states – the question here is: if the system is disturbed slightly from an equilibrium state, will it try to get back to that state or will it move farther away from it? It is usually not good practice to design systems around unstable equilibrium states since the system will always tend to move towards a stable equilibrium condition. Answering the stability question involves a linearization of the equations of motion for small perturbations away from the equilibrium states.

- How the system behaves around a stable equilibrium state – the study of small motions of a mechanical system around a stable equilibrium state lies in the realm of vibrations and leads to predictions of natural frequencies, mode shapes, and damping ratios, each of which is very useful during the design process. The linearized differential equations of motion are used.

- The response to harmonically applied external forces – systems, in stable equilibrium, are often subjected to harmonic external disturbances at known forcing frequencies and their response to these forces provides critical design information. The linearized equations of motion are used.

- The response of the system in the time domain – the fully nonlinear equations of motion are solved numerically to simulate the response of the system to known external forces. Large scale motions and the nonlinear characteristics of system elements are included. This is the numerical equivalent to conducting performance experiments with a prototype of the system. The design information gleaned from these simulations is, perhaps surprisingly to some, not very useful in the early stages of the design. Accurate nonlinear simulations require precise knowledge of system parameters that simply isn't available in the early design phases of a project. The time domain simulations are best left to the prototype testing stage when they can be validated through comparison of predicted and measured system response. Validated time domain simulations are valuable tools to use when considering design changes aimed at improving the measured performance of the system.

The presentation of material in this book divides the practice of engineering dynamics into three parts.

Part 1. Modeling: Deriving Equations of Motion

Dynamic analysis is based on the use of accurate nonlinear equations of motion for a system. Deriving these complicated equations is a task that is prone to error. Because of this, it is important to derive the governing equations twice, using two different methods of analysis, and then prove to yourself that the two sets of equations are the same. This is a time consuming activity but is vital because predictions made from the equations of motion are critical in the design process. Predictions made using equations with errors are not of any use. The first part of the book discusses the generation of nonlinear equations of motion using, firstly, Newton's laws and, secondly, Lagrange's equation. Only when the two methods give the same equations can the analyst proceed to part 2 with confidence.

Part 2. Simulation: Using the Equations of Motion

The second part presents a logical progression of analysis techniques and methods applied to the governing equations of motion for systems. The progression is from equilibrium solutions that find in what states the system would like to be, to analyzing the stability of these equilibrium states (stability is usually considered only in textbooks on control systems but it is vitally important to dynamic systems), to considering small motions about the stable equilibrium states (this topic is covered in textbooks on vibrations but is, again, vital to engineers doing dynamic analysis), to frequency domain analysis (vibrations again), and finally to time domain solutions (these are rarely covered in textbooks).

Part 3. Working with Experimental Data

While not usually considered a part of the design process, analysis of experimental data measured on dynamic systems is critical to creating a successful product. To assist engineers in developing capabilities in this area, part 3 covers the practical use of discrete fourier transforms in analyzing experimental data.

In order to emphasize the idea that any dynamic mechanical system can be analyzed using the sequence of steps presented here, all the exercises at the ends of the chapters are based on 23 mechanical systems defined in an appendix. Any one of these systems could be used as an example of all of the types of dynamic analysis.

This book is based on course notes that I have developed while teaching a one-semester graduate course on dynamics over more than two decades. It could just as well be used in a senior undergraduate dynamics course.

November, 2019

Ronald J. Anderson
Kingston, Canada

About the Companion Website

This book is accompanied by a companion website:

www.wiley.com/go/anderson/engineeringdynamics

The website includes:

- Animations
- Fully worked examples
- Software

Scan this QR code to visit the companion website.

Part I

Modeling: Deriving Equations of Motion

1

Kinematics

Kinematics is defined as the study of motion without reference to the forces that cause the motion. A proper kinematic analysis is an essential first step in any dynamics problem. This is where the analyst defines the degrees of freedom and develops expressions for the absolute velocities and accelerations of the bodies in the system that satisfy all of the physical constraints. The ability to differentiate vectors with respect to time is a critical skill in kinematic analysis.

1.1 Derivatives of Vectors

Vectors have two distinct properties – magnitude and direction. Either or both of these properties may change with time and the time derivative of a vector must account for both.

The rate of change of a vector $\vec{r}$ with respect to time is therefore formed from,

1. The rate of change of magnitude $\left(\frac{d\vec{r}}{dt}\right)_m$.
2. The rate of change of direction $\left(\frac{d\vec{r}}{dt}\right)_d$.

Figure 1.1 shows the vector $\vec{r}(t)$ that changes after a time increment, Δt, to $\vec{r}(t + \Delta t)$.

The difference between $\vec{r}(t)$ and $\vec{r}(t + \Delta t)$ can be defined as the vector $\vec{q}(t)$ shown in Figure 1.1 and, by the rules of vector addition,

$$\vec{r}(t) + \vec{q}(t) = \vec{r}(t + \Delta t) \tag{1.1}$$

or,

$$\vec{q}(t) = \vec{r}(t + \Delta t) - \vec{r}(t). \tag{1.2}$$

Then, using the definition of the time derivative,

$$\frac{d\vec{r}}{dt} = \lim_{\Delta t \to 0} \frac{\vec{r}(t + \Delta t) - \vec{r}(t)}{\Delta t} = \lim_{\Delta t \to 0} \frac{\vec{q}(t)}{\Delta t}. \tag{1.3}$$

Imagine now that Figure 1.1 is compressed to show only an infinitesimally small time interval, Δt. The components of $\vec{q}(t)$ for the interval Δt are shown in Figure 1.1. They are,

1. A component $d\vec{q}_m$ aligned with the vector $\vec{r}$. This is a component that is strictly due to the rate of change of magnitude of $\vec{r}$. The magnitude of $d\vec{q}_m$ is $\dot{r}\Delta t$ where $\dot{r}$ is the rate of

The Practice of Engineering Dynamics, First Edition. Ronald J. Anderson.
© 2020 John Wiley & Sons Ltd. Published 2020 by John Wiley & Sons Ltd.
Companion Website: www.wiley.com/go/anderson/engineeringdynamics

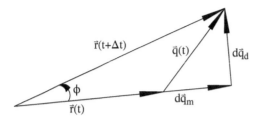

Figure 1.1 A vector changing with time.

change of length (or magnitude) of the vector $\vec{r}$. The direction of $d\vec{q}_m$ is the same as the direction of $\vec{r}$. Let $d\vec{q}_m$ be designated[1] as $\dot{r}\Delta t$.

2. A component $d\vec{q}_d$ that is perpendicular to the vector $\vec{r}$. That is, a component due to the rate of change of direction of the vector. Terms of this type arise only when there is an angular velocity. The rate of change of direction term arises from the time rate of change of the angle ϕ in Figure 1.1 and $\dot{\phi}$ is the magnitude of the angular velocity of the vector. *The rate of change of direction therefore arises from the angular velocity of the vector.* The magnitude of $d\vec{q}_d$ is $\dot{\phi}r\Delta t$ where r is the length of $\vec{r}$. By definition the rate of change of the angle ϕ (i.e. $\dot{\phi}$) has the same positive sense as the angle itself. It is clear that $\dot{\phi}r$ is the "tip speed" one would expect from an object of length r rotating with angular speed $\dot{\phi}$.

The angular velocity is itself a vector quantity since it must specify both the angular speed (i.e. magnitude) and the axis of rotation (i.e. direction). In Figure 1.1, the speed of rotation is $\dot{\phi}$ and the axis of rotation is perpendicular to the page. This results in an angular velocity vector,

$$\vec{\omega} = \dot{\phi}\,\vec{k} \tag{1.4}$$

where the right handed set of unit vectors, $(\vec{i},\vec{j},\vec{k})$, is defined in Figure 1.2. Note that it is essential that right handed coordinate systems be used for dynamic analysis because of the extensive use of the cross product and the directions of vectors arising from it. If there is a right handed coordinate system (x, y, z), with respective unit vectors $(\vec{i},\vec{j},\vec{k})$, then the cross products are such that,

$$\vec{i}\times\vec{j}=\vec{k}$$

$$\vec{j}\times\vec{k}=\vec{i}$$

$$\vec{k}\times\vec{i}=\vec{j}.$$

Using this definition of the angular velocity, the motion of the tip of vector $\vec{r}$, resulting from the angular change in time Δt, can be determined from the cross product

$$(\vec{\omega}\times\vec{r})\Delta t$$

which, by the rules of the vector cross product, has magnitude,

$$|\vec{\omega}||\vec{r}|\Delta t = \dot{\phi}r\Delta t$$

---

1 The convention used here is that a vector with an *overdot* such as $\dot{\vec{r}}$ is used to represent the rate of change of magnitude of the vector and the overdot is not to be interpreted as a shorthand method of signifying the total derivative of the vector. For a scalar function there is only a magnitude and the overdot will represent its rate of change.

Figure 1.2 Even 2D problems are 3D.

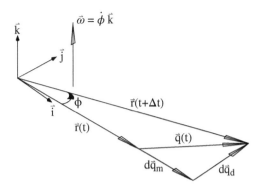

and a direction that, according to the right hand rule[2] used for cross products, is perpendicular to both $\vec{\omega}$ and $\vec{r}$ and, in fact, lies in the direction of $d\vec{q}_d$.

Combining these two terms to get $\vec{q}(t)$ and substituting into Equation 1.3 results in,

$$\frac{d\vec{r}}{dt} = \lim_{\Delta t \to 0} \frac{\dot{\vec{r}}\Delta t + (\vec{\omega} \times \vec{r})\Delta t}{\Delta t} = \dot{\vec{r}} + \vec{\omega} \times \vec{r}. \tag{1.5}$$

The time derivative of any vector, $\vec{r}$, can therefore be written as,

$$\frac{d\vec{r}}{dt} = \underbrace{\dot{\vec{r}}}_{\substack{\text{rate of change} \\ \text{of magnitude} \\ \text{of the vector}}} + \underbrace{\vec{\omega} \times \vec{r}}_{\substack{\text{rate of change} \\ \text{of direction} \\ \text{of the vector}}} \tag{1.6}$$

It is important to understand that the angular velocity vector, $\vec{\omega}$, is the angular velocity of the coordinate system in which the vector, $\vec{r}$, is expressed. There is a danger that the rate of change of direction terms will be included twice if the angular velocity of the vector with respect to the coordinate system in which it measured is used instead. The example presented in Section 1.3 shows a number of different ways to arrive at the derivative of a vector which rotates in a plane.

1.2 Performing Kinematic Analysis

Before proceeding with examples of kinematic analyses we state here the steps that are necessary in achieving a successful result. This first step in any dynamic analysis is vitally important. The goal is to derive expressions for the *absolute velocities and accelerations of the centers of mass* of the bodies making up the system being analyzed. In addition, expressions for the *absolute angular velocities and angular accelerations* of the bodies will be required.

---

2 Let the cross product of two vectors, $\vec{A}$ and $\vec{B}$, be the vector, $\vec{C}$

$$\vec{A} \times \vec{B} = \vec{C}.$$

According to the right hand rule, the direction of $\vec{C}$ is the direction aligned with the thumb of your right hand if you point that hand in the direction of $\vec{A}$ and curl your fingers towards $\vec{B}$.

It is at this first step of the analysis that degrees of freedom are defined and constraints on relative motion between bodies are satisfied.

For this general description of kinematic analysis, we assume that we are analyzing a system that has multiple bodies connected to each other by joints and that we are attempting to derive an expression for the acceleration of the center of mass of a body that is not the first in the assembly.

The procedure is as follows.

1. Find a fixed point (i.e. one having no velocity or acceleration) in the system from which you can begin to write relative position vectors that will lead to the centers of mass of bodies in the system.
2. Define a position vector that goes from the fixed point, through the first body, to the next joint in the system. This is the position of the joint *relative* to the fixed point.
3. Determine how many degrees of freedom, both translational and rotational, are required to define the motion of the relative position vector just defined. The degrees of freedom must be chosen to satisfy the constraints imposed by the joint that connects this body to ground.
4. Define a coordinate system in which the relative position vector will be written and determine the angular velocity of the coordinate system.
5. Repeat the previous three steps as you go from joint to joint in the system, always being careful to satisfy the joint constraints by defining appropriate degrees of freedom.
6. When the desired body is reached, define a final relative position vector from the joint to the center of mass.
7. The sum of all the relative position vectors will be the absolute position of the center of mass and the derivatives of the sum of vectors will yield the absolute velocity and acceleration of the center of mass.

1.3 Two Dimensional Motion with Constant Length

Figure 1.3 shows a rigid rod of length, ℓ, rotating about a fixed point, O, in a plane. An expression for the velocity of the free end of the rod, P, relative to point O is desired.

By definition, the velocity of P relative to O is the time derivative of the position of P relative to O. This position vector is designated $\vec{p}_{P/O}$ and is shown in the figure.

In order to differentiate the position vector, we must have an expression for it and this means we must first choose a coordinate system[3] in which to work. For a start, we can choose a right handed coordinate system fixed in the ground. The set of unit vectors $(\vec{\imath}_0, \vec{\jmath}_0, \vec{k}_0)$ is such a system. The angular velocity of this coordinate system is zero (i.e. $\vec{\omega}_0 = \vec{0}$) since it is fixed in the ground.

An expression for the position of P relative to O in this system is,

$$\vec{p}_{P/O} = \ell \cos\theta \vec{\imath}_0 + \ell \sin\theta \vec{\jmath}_0. \tag{1.7}$$

---

3 Three dimensional sets of unit vectors shown in two dimensional figures such as Figure 1.3 will be shown with the positive sense of the vector out of the plane represented by a curved arrow using the right hand rule (e.g. $\vec{k}_0$ in Figure 1.3).

Figure 1.3 A rigid rod rotating about a fixed point.

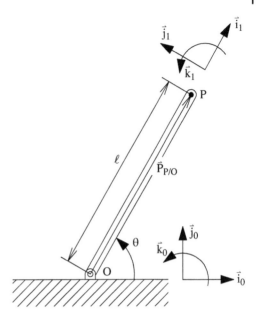

We apply Equation 1.6 to $\vec{p}_{P/O}$ to get,

$$\vec{v}_{P/O} = \frac{d}{dt}(\vec{p}_{P/O}) = \dot{\vec{p}}_{P/O} + \vec{\omega}_0 \times \vec{p}_{P/O}$$

$$= \frac{d}{dt}(\ell \cos\theta \vec{i}_0 + \ell \sin\theta \vec{j}_0) + \vec{0} \times (\ell \cos\theta \vec{i}_0 + \ell \sin\theta \vec{j}_0). \tag{1.8}$$

In this coordinate system, it is clear that there is a rate of change of magnitude of the vector only and the velocity of point P relative to O after performing the simple differentiation is,

$$\vec{v}_{P/O} = -\ell \dot{\theta} \sin\theta \vec{i}_0 + \ell \dot{\theta} \cos\theta \vec{j}_0 = \ell \dot{\theta}(-\sin\theta \vec{i}_0 + \cos\theta \vec{j}_0). \tag{1.9}$$

Another derivation of the velocity of P relative to O might use the system of unit vectors $(\vec{i}_1, \vec{j}_1, \vec{k}_1)$ that are fixed in the rod. The advantage of using this system is that the position vector is easily expressed as,

$$\vec{p}_{P/O} = \ell \vec{i}_1. \tag{1.10}$$

Note that the length of this vector is a constant so that the total derivative must come from its rate of change of direction. The angular velocity of the coordinate system is equal to the angular velocity of the rod since the coordinate system is fixed in the rod. That is,

$$\vec{\omega}_1 = \dot{\theta} \vec{k}_1 \tag{1.11}$$

and the velocity of P relative to O is therefore[4],

$$\vec{v}_{P/O} = \frac{d}{dt}(\vec{p}_{P/O}) = \dot{\vec{p}}_{P/O} + \vec{\omega}_1 \times \vec{p}_{P/O} = (\dot{\ell} \vec{i}_1) + (\dot{\theta} \vec{k}_1) \times (\ell \vec{i}_1). \tag{1.12}$$

Since ℓ is constant, $\dot{\ell} = 0$, and the final result is,

$$\vec{v}_{P/O} = (\dot{\theta} \vec{k}_1) \times (\ell \vec{i}_1) = \ell \dot{\theta} \vec{j}_1. \tag{1.13}$$

---

4 Readers are encouraged to review the rules for cross multiplication of vectors.

We now have two expressions for $\vec{v}_{P/O}$ (Equations 1.9 and 1.13). Since there can only be one value of this relative velocity, the two expressions must be equal to each other. However, the use of two different coordinate systems makes them look different. In order to compare them, we must be able to transform results from one coordinate system to the other.

Keep in mind that sequential sets of unit vectors are related to each other by simple plane rotations. Also note that the unit vectors are not related to any point in the system – they simply express directions. Given these two facts, we can relate the two sets of unit vectors we have been using by noting that $\vec{k}_1 = \vec{k}_0$ (i.e. the plane rotation relating the two sets is a rotation about the $\vec{k}_0$ or $\vec{k}_1$ axis). The relationships between the two sets of unit vectors can be expressed as follows.

$$\begin{Bmatrix} \vec{\imath}_1 \\ \vec{\jmath}_1 \\ \vec{k}_1 \end{Bmatrix} = \begin{bmatrix} \cos\theta & \sin\theta & 0 \\ -\sin\theta & \cos\theta & 0 \\ 0 & 0 & 1 \end{bmatrix} \begin{Bmatrix} \vec{\imath}_0 \\ \vec{\jmath}_0 \\ \vec{k}_0 \end{Bmatrix} \tag{1.14}$$

or

$$\begin{Bmatrix} \vec{\imath}_0 \\ \vec{\jmath}_0 \\ \vec{k}_0 \end{Bmatrix} = \begin{bmatrix} \cos\theta & -\sin\theta & 0 \\ \sin\theta & \cos\theta & 0 \\ 0 & 0 & 1 \end{bmatrix} \begin{Bmatrix} \vec{\imath}_1 \\ \vec{\jmath}_1 \\ \vec{k}_1 \end{Bmatrix}. \tag{1.15}$$

The first transformation (Equation 1.14) can be used with Equation 1.13 to show that,

$$\vec{v}_{P/O} = \ell\dot{\theta}\vec{\jmath}_1 = \ell\dot{\theta}(-\sin\theta\vec{\imath}_0 + \cos\theta\vec{\jmath}_0). \tag{1.16}$$

This is the same result as that shown in Equation 1.9.

Similarly, the second transformation (Equation 1.15) can be used with Equation 1.9 to show that,

$$\begin{aligned} \vec{v}_{P/O} &= -\ell\dot{\theta}\sin\theta\vec{\imath}_0 + \ell\dot{\theta}\cos\theta\vec{\jmath}_0 \\ &= -\ell\dot{\theta}\sin\theta(\cos\theta\vec{\imath}_1 - \sin\theta\vec{\jmath}_1) + \ell\dot{\theta}\cos\theta(\sin\theta\vec{\imath}_1 + \cos\theta\vec{\jmath}_1). \end{aligned} \tag{1.17}$$

Expanding this yields,

$$\vec{v}_{P/O} = \ell\dot{\theta}(-\sin\theta\cos\theta\vec{\imath}_1 + \sin^2\theta\vec{\jmath}_1 + \cos\theta\sin\theta\vec{\imath}_1 + \cos^2\theta\vec{\jmath}_1) = \ell\dot{\theta}\vec{\jmath}_1. \tag{1.18}$$

This is the same result as that shown in Equation 1.13.

The transformations described in this example are typical of those used in dynamic analysis. Dynamicists are prone to using whatever coordinate system is appropriate at the time and, sometimes, there are many intermediate coordinate systems used in deriving the final system. Nevertheless, each coordinate system in the sequence must be right handed and must be generated by a simple plane rotation from the preceding system.

1.4 Two Dimensional Motion with Variable Length

Figure 1.4 shows a rigid body rotating in a plane about a fixed point O. The body has a slot cut in it and a small object A slides in this slot. Expressions for the velocity and acceleration of A are desired.

Figure 1.4 A slider in a slot.

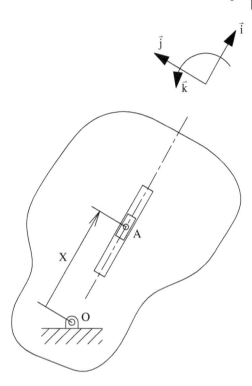

Since the distance from O to A changes with time, we start by defining a variable distance x from O to A. A set of rotating unit vectors $(\vec{i}, \vec{j}, \vec{k})$, fixed in the rotating body, as shown, is appropriate for this analysis since the position vector $\vec{p}_{A/O}$ is aligned with $\vec{i}$, thereby making it easy to write.

The angular velocity of the body is not specified in magnitude but the fact that the body rotates in a plane fixes the direction of the angular velocity to be $\vec{k}$. We assume that the angular velocity is,

$$\vec{\omega} = \omega\vec{k}$$

where ω is not constant so that $\dot{\omega}$ (i.e. the rate of change of magnitude of the angular velocity vector) exists.

The position of A with respect to O is then,

$$\vec{p}_{A/O} = x\vec{i}$$

and, differentiating this, we find the velocity of A with respect to O to be,

$$\vec{v}_{A/O} = \frac{\mathrm{d}}{\mathrm{d}t}(\vec{p}_{A/O}) = \frac{\mathrm{d}}{\mathrm{d}t}(x\vec{i}) = \dot{x}\vec{i} + (\omega\vec{k}) \times (x\vec{i}) = \dot{x}\vec{i} + \omega x\vec{j}. \tag{1.19}$$

The acceleration of the slider relative to point O is defined to be,

$$\vec{a}_{A/O} = \frac{\mathrm{d}}{\mathrm{d}t}(\vec{v}_{A/O})$$

or

$$\vec{a}_{A/O} = \frac{d}{dt}(\dot{x}\vec{i} + \omega x\vec{j}) = (\ddot{x}\vec{i} + \dot{\omega}x\vec{j} + \omega\dot{x}\vec{j}) + (\omega\vec{k}) \times (\dot{x}\vec{i} + \omega x\vec{j})$$

$$= (\ddot{x} - \omega^2 x)\vec{i} + (\dot{\omega}x + 2\omega\dot{x})\vec{j}. \tag{1.20}$$

Since both the velocity $\vec{v}_{A/O}$ and the acceleration $\vec{a}_{A/O}$ are relative to the fixed or inertial point O, they are in fact the absolute velocity and acceleration of point A. We commonly write absolute velocities and accelerations without subscripts yielding,

$$\vec{v}_A = \dot{x}\vec{i} + \omega x\vec{j} \tag{1.21}$$

and

$$\vec{a}_A = (\ddot{x} - \omega^2 x)\vec{i} + (\dot{\omega}x + 2\omega\dot{x})\vec{j}. \tag{1.22}$$

1.5 Three Dimensional Kinematics

Figure 1.5 shows a three degree of freedom robot. The horizontal arm AB is of fixed length ℓ and is free to rotate about a vertical axis through point A with angular speed ω_0. Arm BC has a variable length x and is free to rotate about an axis passing through points A and B with angular speed ω_1. The end effector is located at point C. Of interest for the kinematic analysis are expressions for the absolute velocity and acceleration of the end effector.

As a first step we define the right handed coordinate system $(\vec{i}_0, \vec{j}_0, \vec{k}_0)$ fixed in the arm AB. This is a rotating coordinate system with angular velocity $\vec{\omega}_0 = \omega_0 \vec{k}_0$.

The process of finding the absolute velocity and acceleration of point C is just as outlined in Section 1.2. The first step is to find a fixed point. In this system, point A serves the purpose as it has no velocity or acceleration.

The next step is to define a position vector that goes from A to C. This is the vector $\vec{P}_{C/A}$ shown in Figure 1.6. Notice that it goes through space from A to C and its rate of change cannot be described directly using the motions of the physical components of the system.

We must, in fact, work with relative position vectors that go from the fixed point to the point of interest by passing from joint to joint. In this case, we define first a vector that goes from A to B ($\vec{P}_{B/A}$) and then add to it a vector that goes from B to C ($\vec{P}_{C/B}$). That is,

$$\vec{P}_{C/A} = \vec{P}_{B/A} + \vec{P}_{C/B}. \tag{1.23}$$

Figure 1.5 A three dimensional robot.

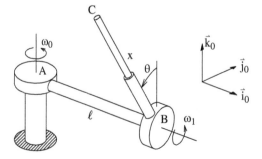

Figure 1.6 Relative position vectors.

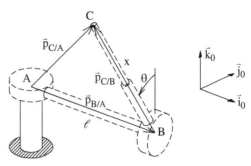

By definition, the absolute velocity of C is the time rate of change of its position with respect to a fixed point. That is,

$$\vec{v}_C = \frac{d}{dt}(\vec{p}_{C/A}) \tag{1.24}$$

which, upon substitution of Equation 1.23, becomes,

$$\vec{v}_C = \frac{d}{dt}(\vec{p}_{B/A}) + \frac{d}{dt}(\vec{p}_{C/B}) \tag{1.25}$$

which in turn becomes,

$$\vec{v}_C = \vec{v}_{B/A} + \vec{v}_{C/B} \tag{1.26}$$

where we see that the absolute velocity of C can be expressed as the sum of the velocities of points in the vector chain relative to previous points in the chain so long as the first point is stationary.

Simply differentiating Equation 1.26 with respect to time gives the corresponding expression for accelerations.

$$\vec{a}_C = \vec{a}_{B/A} + \vec{a}_{C/B}. \tag{1.27}$$

With respect to the particular system being considered here, we can write,

$$\vec{p}_{B/A} = \ell \, \vec{\imath}_0 \tag{1.28}$$

and,

$$\vec{p}_{C/B} = -x \sin \theta \, \vec{\jmath}_0 + x \cos \theta \, \vec{k}_0. \tag{1.29}$$

Considering first the position of B with respect to A we see that the length ℓ is constant so there will be no rate of change of magnitude of the vector but there will be a rate of change of direction since the coordinate system is rotating. We find,

$$\vec{v}_{B/A} = \left(\omega_0 \, \vec{k}_0 \right) \times (\ell \, \vec{\imath}_0) = \omega_0 \, \ell \, \vec{\jmath}_0. \tag{1.30}$$

We differentiate again, noting that ω_0 is not constant so that there will be a rate of change of magnitude this time, and find,

$$\vec{a}_{B/A} = (\dot{\omega}_0 \, \ell \, \vec{\jmath}_0) + \left(\omega_0 \, \vec{k}_0 \right) \times (\omega_0 \, \ell \, \vec{\jmath}_0)$$
$$= \dot{\omega}_0 \, \ell \, \vec{\jmath}_0 - \omega_0^2 \, \ell \, \vec{\imath}_0. \tag{1.31}$$

The rate of change of $\vec{p}_{C/B}$ is a little more complicated for two reasons. First, the vector is not in the body to which the coordinate system being used is fixed so there will be a rate of change of magnitude arising from the time derivatives of the trigonometric functions. Second, the vector itself is not of fixed length so terms involving the rate of change of x will appear. Differentiating yields,

$$\vec{v}_{C/B} = \left(-\dot{x}\sin\theta - x\dot{\theta}\cos\theta\right)\vec{j}_0$$
$$+ \left(\dot{x}\cos\theta - x\dot{\theta}\sin\theta\right)\vec{k}_0$$
$$+ \left(\omega_0\,\vec{k}_0\right) \times \left(-x\sin\theta\,\vec{j}_0 + x\cos\theta\,\vec{k}_0\right) \tag{1.32}$$

which, noting that $\dot{\theta} = \omega_1$ (see Figure 1.5), expands to,

$$\vec{v}_{C/B} = \left(x\omega_0\sin\theta\right)\vec{i}_0 + \left(-\dot{x}\sin\theta - x\omega_1\cos\theta\right)\vec{j}_0$$
$$+ \left(\dot{x}\cos\theta - x\omega_1\sin\theta\right)\vec{k}_0. \tag{1.33}$$

Finally, we differentiate $\vec{v}_{C/B}$ to get $\vec{a}_{C/B}$ as follows.

$$\vec{a}_{C/B} = \left(x\dot{\omega}_0\sin\theta + \dot{x}\omega_0\sin\theta + x\omega_0\dot{\theta}\cos\theta\right)\vec{i}_0$$
$$+ \left(-\ddot{x}\sin\theta - \dot{x}\dot{\theta}\cos\theta - \dot{x}\omega_1\cos\theta - x\dot{\omega}_1\cos\theta + x\omega_1\dot{\theta}\sin\theta\right)\vec{j}_0$$
$$+ \left(\ddot{x}\cos\theta - \dot{x}\dot{\theta}\sin\theta - \dot{x}\omega_1\sin\theta - x\dot{\omega}_1\sin\theta - x\omega_1\dot{\theta}\cos\theta\right)\vec{k}_0$$
$$+ \left(\omega_0\,\vec{k}_0\right) \times \vec{v}_{C/B} \tag{1.34}$$

which, after considerable effort and again noting that $\dot{\theta} = \omega_1$, expands to,

$$\vec{a}_{C/B} = \left(x\,\dot{\omega}_0\sin\theta + 2\,\dot{x}\,\omega_0\sin\theta + 2\,x\,\omega_0\,\omega_1\cos\theta\right)\vec{i}_0$$
$$+ \left(-\ddot{x}\sin\theta - 2\,\dot{x}\,\omega_1\cos\theta - x\,\dot{\omega}_1\cos\theta + x(\omega_0^2 + \omega_1^2)\sin\theta\right)\vec{j}_0$$
$$+ \left(\ddot{x}\cos\theta - 2\,\dot{x}\,\omega_1\sin\theta - x\,\dot{\omega}_1\sin\theta - x\,\omega_1^2\cos\theta\right)\vec{k}_0. \tag{1.35}$$

If you have worked through the derivation of Equation 1.35 you will be aware that the probability of making a mistake when deriving equations such as this is high. A quick, approximate check on the accuracy of your work can be made by verifying that every term in the acceleration expression has dimensions of acceleration or, more simply, contains two derivatives of displacement variables. That is, terms like $\ddot{x}$ are obviously accelerations whereas terms like $x\,\dot{\omega}_0$ might require a little thought before realizing that $\dot{\omega}_0$ is the second derivative of an angle and must be scaled by a length, in this case x, in order to be a translational acceleration. Products of angular velocities such as $x\,\omega_0\,\omega_1$ and $x\,\omega_1^2$ have two derivatives of angles multiplied together and are again scaled by a length, x, to become translational accelerations. In addition, it is somewhat comforting to see several terms that have the Coriolis factor of 2 associated with them. Suspicion should be raised when factors other than 1 or 2 are seen in acceleration expressions.

We now simply add the relative acceleration vectors to arrive at the absolute acceleration of point C. That is,

$$\vec{a}_C = \vec{a}_A + \vec{a}_{B/A} + \vec{a}_{C/B}$$
$$= \left(-\omega_0^2\,\ell + x\,\dot{\omega}_0\sin\theta + 2\,\dot{x}\,\omega_0\sin\theta + 2\,x\,\omega_0\,\omega_1\cos\theta\right)\vec{i}_0$$
$$+ \left(\dot{\omega}_0\,\ell - \ddot{x}\sin\theta - 2\,\dot{x}\,\omega_1\cos\theta - x\,\dot{\omega}_1\cos\theta + x\left(\omega_0^2 + \omega_1^2\right)\sin\theta\right)\vec{j}_0$$
$$+ \left(\ddot{x}\cos\theta - 2\,\dot{x}\,\omega_1\sin\theta - x\,\dot{\omega}_1\sin\theta - x\,\omega_1^2\cos\theta\right)\vec{k}_0. \tag{1.36}$$

1.6   Absolute Angular Velocity and Acceleration

When working with three dimensional dynamic systems it is important to have an expression for the *absolute angular velocity vector* for a rigid body in order to be able to write an expression for its angular momentum vector. The angular momentum is required for moment balances.

Relative angular velocity vectors can be added together in the same way that relative velocity vectors were in Section 1.5. That is, having established the angular velocity of one body in a chain of bodies with respect to a stationary body (i.e. the absolute angular velocity of the body), we simply go through the chain adding the relative angular velocity of neighboring bodies as we pass through the joints connecting them.

For example, the absolute angular velocity of body *BC* in Figure 1.5 can be determined as follows,

$$\vec{\omega}_{BC} = \vec{\omega}_{AB} + \vec{\omega}_{BC/AB} \tag{1.37}$$

where the joint at *A* constrains *AB* to rotate about a vertical axis relative to the ground, so that,

$$\vec{\omega}_{AB} = \omega_0 \, \vec{k}_0 \tag{1.38}$$

is the absolute angular velocity of *AB*. The joint at *B* constrains *BC* to rotate about the axis of *AB* with an angular velocity that is *relative* to *AB* giving,

$$\vec{\omega}_{BC/AB} = \omega_1 \, \vec{\imath}_0. \tag{1.39}$$

Substituting Equations 1.38 and 1.39 into Equation 1.37 gives the absolute angular velocity of *BC*

$$\vec{\omega}_{BC} = \omega_1 \, \vec{\imath}_0 + \omega_0 \, \vec{k}_0. \tag{1.40}$$

The *absolute angular acceleration* of *BC* is, by definition, the time rate of change of the *absolute angular velocity vector* of *BC*. In this example we note that the angular velocity vector is expressed in a rotating coordinate system so that there will be both a rate of change of magnitude and a rate of change of direction. The coordinate system has angular velocity $\vec{\omega}_0 = \omega_0 \, \vec{k}_0$. Using the symbol α for angular acceleration we can write,

$$\vec{\alpha}_{BC} = \frac{d}{dt}(\omega_1 \, \vec{\imath}_0 + \omega_0 \, \vec{k}_0) \tag{1.41}$$

which becomes, upon differentiation,

$$\vec{\alpha}_{BC} = (\dot{\omega}_1 \, \vec{\imath}_0 + \dot{\omega}_0 \, \vec{k}_0) + (\omega_0 \, \vec{k}_0) \times (\omega_1 \, \vec{\imath}_0 + \omega_0 \, \vec{k}_0). \tag{1.42}$$

The final result, after performing the cross-multiplication in Equation 1.42, is that the absolute angular acceleration of *BC* is,

$$\vec{\alpha}_{BC} = \dot{\omega}_1 \, \vec{\imath}_0 + \omega_0 \, \omega_1 \, \vec{\jmath}_0 + \dot{\omega}_0 \, \vec{k}_0. \tag{1.43}$$

1.7 The General Acceleration Expression

In Section 1.4 we derived an acceleration expression for a very specific example. The final result (shown in Equation 1.20) has an interesting and, perhaps, unexpected form. In particular, the origin of the term that has twice the product of an angular velocity and a translational velocity (i.e. $2\omega\dot{x}$) is not immediately obvious. The origin of all of the acceleration terms in a general expression like that in Equation 1.20 is described below. The description is offered twice – first in a mathematical form then in a graphical form.

In general, the derivation of an expression for the acceleration of a point (say P) relative to another point (say O) starts with the position vector of P with respect to O and then differentiates it twice. Each differentiation must take account of the angular velocity of the coordinate system being used to express the vectors.

Let the position vector be

$$\vec{P}_{P/O}. \tag{1.44}$$

Then, applying Equation 1.6, the velocity is,

$$\vec{v}_{P/O} = \underbrace{\dot{\vec{P}}_{P/O}}_{\text{radial}} + \underbrace{\vec{\omega} \times \vec{P}_{P/O}}_{\text{tangential}} \tag{1.45}$$

where the directions of the two components are defined. The rate of change of magnitude term is aligned with the position vector and is thus termed *radial* and the rate of change of direction component is perpendicular to the position vector and is therefore *tangential*.

Differentiating the velocity expression of Equation 1.45 yields,

$$\vec{a}_{P/O} = \frac{\mathrm{d}}{\mathrm{d}t}(\dot{\vec{P}}_{P/O}) + \frac{\mathrm{d}}{\mathrm{d}t}(\vec{\omega} \times \vec{P}_{P/O})$$
$$= \frac{\mathrm{d}}{\mathrm{d}t}(\dot{\vec{P}}_{P/O}) + \frac{\mathrm{d}}{\mathrm{d}t}(\vec{\omega}) \times \vec{P}_{P/O} + \vec{\omega} \times \frac{\mathrm{d}}{\mathrm{d}t}(\vec{P}_{P/O}) \tag{1.46}$$

and, applying Equation 1.6 to Equation 1.46, we can write,

$$\vec{a}_{P/O} = (\ddot{\vec{P}}_{P/O} + \vec{\omega} \times \dot{\vec{P}}_{P/O}) + \frac{\mathrm{d}\vec{\omega}}{\mathrm{d}t} \times \vec{P}_{P/O} + \vec{\omega} \times (\dot{\vec{P}}_{P/O} + \vec{\omega} \times \vec{P}_{P/O}). \tag{1.47}$$

After collecting terms and substituting $\vec{\alpha}$ (the angular acceleration vector) for $\dfrac{\mathrm{d}\vec{\omega}}{\mathrm{d}t}$, we get the well-known result,

$$\vec{a}_{P/O} = \underbrace{\ddot{\vec{P}}_{P/O}}_{\text{radial}} + \underbrace{\vec{\alpha} \times \vec{P}_{P/O}}_{\text{tangential}} + \underbrace{2\vec{\omega} \times \dot{\vec{P}}_{P/O}}_{\text{Coriolis}} + \underbrace{\vec{\omega} \times (\vec{\omega} \times \vec{P}_{P/O})}_{\text{centripetal}}. \tag{1.48}$$

Each of the terms in Equation 1.48 has a name and a physical meaning, as follows.

1. $\ddot{\vec{P}}_{P/O}$ is the *radial acceleration*. This is nothing more than the second derivative of the distance between O and P and it is aligned with $\vec{P}_{P/O}$.

2. $\vec{\alpha} \times \vec{P}_{P/O}$ is the *tangential acceleration*. It is called the tangential acceleration because it is aligned with the direction in which the point P would move if it were a fixed distance from O and were rotating about O (i.e. in a direction perpendicular to a line passing through O and P). Notice that $\vec{\alpha}$ is the total derivative of the angular velocity including

its rate of change of magnitude and its rate of change of direction. As a result, $\vec{a}$ may not be aligned with $\vec{\omega}$.

3. $2\vec{\omega} \times \dot{\vec{P}}_{P/O}$ is the *Coriolis acceleration*[5]. The vectorial approach to finding the Coriolis acceleration is in many ways preferable to the scalar approach put forward in many books on dynamics. The magnitude of the radial velocity $\dot{\vec{P}}_{P/O}$ is often referred to in reference books as v_{rel} and the Coriolis acceleration is seen written as $2\omega v_{\text{rel}}$ where the reader is left to determine its direction from a complicated set of rules. Consideration of Equations 1.45–1.47 shows that there are two very different types of terms that combine to form the Coriolis acceleration with its remarkable 2. The two terms are equal in magnitude and direction (i.e. each is $\vec{\omega} \times \dot{\vec{P}}_{P/O}$). One of these arises from part of the rate of change of magnitude of the tangential velocity of P. The second arises from the rate of change of direction of the radial velocity of P.

4. $\vec{\omega} \times (\vec{\omega} \times \vec{P}_{P/O})$ is the *centripetal acceleration*. In 2D circular motion. this is commonly written as $\omega^2 r$ and points toward the center of the circle. For the general points O and P used here, the centripetal acceleration points from P to O.

It is possible to visualize the acceleration components using a simple graphical construction. As an example, we can use the slider in a slot system shown in Figure 1.4 for which we have already derived both the velocity (Equation 1.19) and the acceleration (Equation 1.20) in body fixed coordinates.

Remember that rates of change of magnitude are aligned with the vector that is changing and rates of change of direction are perpendicular to the original vector and are pointed in the direction that the tip of the vector would move if it had the prescribed angular velocity and were simply rotating about its tail.

Figure 1.7 shows the two components of the velocity of the slider in the inner circle. A component is labeled with a Δm to indicate that it results from a rate of change of magnitude or a Δd to show that it results from a rate of change of direction. The two terms here are the radial velocity $\dot{x}$ aligned with the original position vector (a rate of change of magnitude term) and the tangential velocity ωx (a rate of change of direction term) perpendicular to the position vector, pointing in the direction that the angular velocity would cause point A to move if it were simply rotating about O, and with magnitude equal to the vector length multiplied by the angular speed. These are the same two terms that appear in Equation 1.19.

Between the inner and outer circles on Figure 1.7 are the components of the acceleration. To get these terms, we treat the velocity components as separate vectors that can change in magnitude and direction and apply to each of them the same procedure we used on the position vector in the preceding paragraph.

Consider first the radial velocity ($\dot{x}$). Its rate of change of magnitude will be aligned with it and will be equal to the rate of change of its length (i.e. the derivative of $\dot{x}$ with respect to time or $\ddot{x}$). As we watch it rotate about its tail with angular speed ω, the tip of this vector will move at the rate $\omega \dot{x}$ in the direction $\vec{j}$ shown on the figure. This rate of change of direction of the radial velocity is one half of the Coriolis acceleration.

---

5 Gaspard Gustave de Coriolis (1792–1843), an engineer and mathematician, introduced the terms "work" and "kinetic energy" to engineering analysis but is best remembered for showing that the laws of motion could be used in a rotating reference frame if an extra term called the Coriolis acceleration is added to the equations of motion.

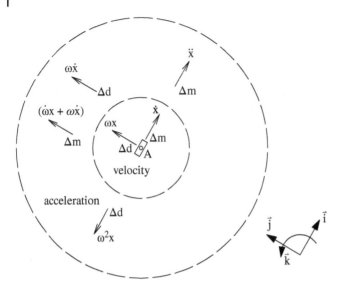

Figure 1.7 The velocity and acceleration components of the slider.

Next consider the tangential velocity (ωx). Its rate of change of magnitude is aligned with it and consists of the time derivative of ωx which has two terms by the chain rule of differentiation (i.e. $\dot{\omega} x$ and $\omega \dot{x}$). Notice that the second of these terms is the other half of the Coriolis acceleration. The rate of change of direction of the tangential velocity is found by rotating it about its tail with angular speed ω and finding that it has magnitude $\omega(\omega x)$ or $\omega^2 x$ and points from A to O. The rate of change of direction of the tangential velocity is, in fact, the centripetal acceleration.

The vector sum of these components yields the same acceleration as Equation 1.20.

Exercises

Descriptions of the systems referred to in the exercises are contained in Appendix A.

1.1 Show that the absolute velocity of point B of system 1 is,

$$\vec{v}_B = [-d_2(\dot{\theta}_1 + \dot{\theta}_2) \sin \theta_2]\vec{i} + [d_1\dot{\theta}_1 + d_2(\dot{\theta}_1 + \dot{\theta}_2) \cos \theta_2]\vec{j}.$$

1.2 Show that the absolute acceleration of point B of system 1 is,

$$\vec{a}_B = [-d_1\dot{\theta}_1^2 - d_2(\ddot{\theta}_1 + \ddot{\theta}_2) \sin \theta_2 - d_2(\dot{\theta}_1 + \dot{\theta}_2)^2 \cos \theta_2]\vec{i}$$
$$+ [d_1\ddot{\theta}_1 + d_2(\ddot{\theta}_1 + \ddot{\theta}_2) \cos \theta_2 - d_2(\dot{\theta}_1 + \dot{\theta}_2)^2 \sin \theta_2]\vec{j}. \qquad (1.49)$$

1.3 Let the coordinate system shown in system 2 be fixed in AB and show that the acceleration of C with respect to B is,

$$\vec{a}_{C/B} = [-d(\ddot{\theta} - \ddot{\phi}) \cos \phi - d(\dot{\theta} - \dot{\phi})^2 \sin \phi]\vec{i}$$
$$+ [d(\ddot{\theta} - \ddot{\phi}) \sin \phi - d(\dot{\theta} - \dot{\phi})^2 \cos \phi]\vec{j}. \qquad (1.50)$$

1.4 Let the coordinate system shown in system 2 be fixed in *BC* and repeat the previous problem. Note that the angle ϕ is constant in this coordinate system so that the vector $\vec{\rho}_{C/B}$ has no rate of change of magnitude.

1.5 Show that the "rolls without slipping condition" in system 5 requires that the drum have an angular velocity $\omega = \dot{\theta} + \dot{x}/r$ in the counterclockwise direction.

1.6 Show that the absolute acceleration of the center of the drum in system 5 has a component,

$$\ddot{x} + r\ddot{\theta} + (d - x)\dot{\theta}^2$$

down the plane and a component perpendicular to the plane that can be written as,

$$-(d - x)\ddot{\theta} + r\dot{\theta}^2 + 2\dot{x}\dot{\theta}.$$

1.7 Define a right handed coordinate system in system 6 where $\vec{i}$ is aligned with *OA* and $\vec{k}$ points upward and show that,

$$\begin{aligned}
\vec{a}_m &= [r\ddot{\theta}\sin\theta + r\dot{\theta}^2\cos\theta - \omega^2(d - r\cos\theta)]\vec{i} \\
&+ [\dot{\omega}(d - r\cos\theta) + 2\omega r\dot{\theta}\sin\theta]\vec{j} \\
&+ [r\ddot{\theta}\cos\theta - r\dot{\theta}^2\sin\theta]\vec{k}.
\end{aligned}$$

1.8 Show that the "rolls without slipping condition" in system 14 requires that the disk have an angular velocity $\omega = \dot{\theta}(R + r)/r$ in the clockwise direction.

1.9 Show that the acceleration of the center of mass of the uniform rod in system 16 has a component,

$$r\ddot{\theta}\sin\phi - r\dot{\theta}^2\cos\phi - \ell(\dot{\theta} + \dot{\phi})^2$$

aligned with the rod and a component perpendicular to the rod that can be written as,

$$r\ddot{\theta}\cos\phi + r\dot{\theta}^2\sin\phi + \ell(\ddot{\theta} + \ddot{\phi}).$$

1.10 Using the coordinate system of Exercise 1.7, show that the absolute angular acceleration of the massless rigid rod in system 6 is,

$$\vec{\alpha}_{Am} = -\omega\,\dot{\theta}\,\vec{i} + \ddot{\theta}\,\vec{j} + \dot{\omega}\,\vec{k}.$$

1.11 Using a coordinate system fixed in the ground with $\vec{i}$ positive to the right and $\vec{j}$ positive up, show that the acceleration of the center of mass of the the rod *OA* in system 12 is

$$\vec{a}_G = (d\ddot{\theta}\sin\theta + d\dot{\theta}^2\cos\theta)\vec{i} + (-d\ddot{\theta}\cos\theta + d\dot{\theta}^2\sin\theta)\vec{j}.$$

1.12 Show that the absolute acceleration of the center of mass of the rod in system 16 is,

$$\vec{a}_G = [-r\dot{\theta}^2 - \ell(\ddot{\theta} + \ddot{\phi})\sin\phi - \ell(\dot{\theta} + \dot{\phi})^2 \cos\phi]\vec{i}$$
$$+ [r\ddot{\theta} + \ell(\ddot{\theta} + \ddot{\phi})\cos\phi - \ell(\dot{\theta} + \dot{\phi})^2 \sin\phi]\vec{j}$$

where $\vec{i}$ points from O to A.

1.13 Consider system 23. Using the rotating coordinate system shown, show that the absolute velocity of the mass is,

$$\vec{v}_m = (\dot{\ell}\sin\phi + \ell\dot{\phi}\cos\phi)\vec{i} + (\ell\dot{\theta}\sin\phi)\vec{j} + (-\dot{\ell}\cos\phi + \ell\dot{\phi}\sin\phi)\vec{k}.$$

1.14 Consider system 23 again. Define a ground fixed coordinate system $(\vec{i}_0, \vec{j}_0, \vec{k}_0)$ obtained by a plane rotation of the $(\vec{i}, \vec{j}, \vec{k})$ system through an angle θ in the negative $\vec{k}$ direction. That is, $\vec{i}_0$ points from B to A and $\vec{k}_0$ is aligned with $\vec{k}$. Show that the position of the mass with respect to the point O in this system is,

$$\vec{p}_{m/O} = (\ell\sin\phi\cos\theta)\vec{i}_0 + (\ell\sin\phi\sin\theta)\vec{j}_0 + (-\ell\cos\phi)\vec{k}_0.$$

Differentiate this position vector to get the absolute velocity of the mass and show that you could get the same result by transforming the result of Exercise 1.13 using a plane rotation.

1.15 Finally, for system 23. Define a body fixed coordinate system $(\vec{i}_1, \vec{j}_1, \vec{k}_1)$ obtained by a plane rotation of the $(\vec{i}, \vec{j}, \vec{k})$ system through an angle ϕ in the negative $\vec{j}$ direction. That is, $\vec{k}_1$ points from m to O and $\vec{j}_1$ is aligned with $\vec{j}$. The position of the mass with respect to the point O in this system is,

$$\vec{p}_{m/O} = -\ell\vec{k}_1.$$

Differentiate this position vector to get the absolute velocity of the mass and show that you could get the same result by transforming the result of Exercise 1.13 using a plane rotation. Be sure to get the correct angular velocity for the coordinate system before differentiating.

2

Newton's Equations of Motion

2.1 The Study of Motion

The study of motion has a long history. Much of the early work was motivated by a desire to study the motions of celestial bodies and the early scientists who worked in this area have contributed to the body of knowledge we rely upon when we analyze the motion of mechanical systems.

Galileo Galilei (1564–1642) was the first person to describe the concept of *inertia*. He stated that inertia was the property of matter that causes it to resist changing its state of motion unless compelled to by applied forces. This idea is fundamental to the field of dynamics.

Issac Newton (1642–1727) developed the three laws of motion that form the basis for our study of dynamics. His first and third laws are basically a restatement of Galileo's concept of inertia and forces of interaction. His second law, however, formally introduces the fact that the effect of a force on an object is to cause a rate of change of its momentum that equals the force both in magnitude and direction. This is the fundamental building block upon which all of the material presented here is based.

2.2 Newton's Laws

Newton developed three laws of motion that are, for the most part, common sense. He did however introduce the concept of "mass", which is something that remains with us today, although his original definition has been somewhat changed by the work of Einstein et al. Nevertheless, the majority of problems in mechanical engineering can still be handled by reference to Newton's laws. They are,

1. Every body perseveres in its state of rest or of uniform motion in a straight line, except in-so-far-as it is compelled to change that state by impressed forces.
2. The rate of change of momentum of a body is equal to the impressed force and takes place in the direction of the straight line along which the force acts.
   This is a very interesting statement. It relates two vector quantities (momentum and force) and says that the momentum vector has a rate of change (derivative) that is

The Practice of Engineering Dynamics, First Edition. Ronald J. Anderson.
© 2020 John Wiley & Sons Ltd. Published 2020 by John Wiley & Sons Ltd.
Companion Website: www.wiley.com/go/anderson/engineeringdynamics

equal to the applied (impressed) force. This is Newton's famous second law that is often stated as $F = ma$. If you performed experiments and plotted the applied force versus the resulting acceleration you would see that the curves were straight lines and would conclude that the acceleration was proportional to the applied force. The constant of proportionality is the "mass" and, to this day, we are unable to measure mass directly. We can measure force because of the way in which it causes materials to deflect but mass can only be inferred from a force measurement . There are four basic quantities related to the study of dynamics – force, mass, length, and time. Of these, we can only measure three – force, length, and time. This fact becomes important when we choose a system of units for analysis.

3. Reaction is always equal and opposite to action, that is to say, the actions of two bodies upon each other are always equal and directly opposite.

Newton's three laws, being simple and intuitive, are the foundation upon which most of the field of dynamics is built.

2.3 Newton's Second Law for a Particle

The second law can be stated for a single particle as,

$$\vec{F} = \frac{d\vec{G}}{dt} \qquad (2.1)$$

where $\vec{F}$ is the total externally applied force acting on the particle. $\vec{G} = m\vec{v}$ is the linear momentum of the particle. m is the mass of the particle. $\vec{v}$ is the absolute velocity [1].

Immediately upon considering applications of Equation 2.1, it becomes apparent that an ability to work with derivatives of vectors is required. The second law makes specific reference to the time rate of change of the linear momentum vector $\frac{d\vec{G}}{dt}$. Substituting $\vec{G} = m\vec{v}$ into Equation 2.1 yields,

$$\vec{F} = \frac{d}{dt}(m\vec{v}) = \frac{dm}{dt}\vec{v} + m\frac{d\vec{v}}{dt}. \qquad (2.2)$$

The first term in the expansion (i.e. $\frac{dm}{dt}\vec{v}$) is non-zero only when systems with variable mass are considered. Rockets, for instance, have a fairly large rate of change of mass as fuel is consumed during takeoff and $\frac{dm}{dt}$ must be considered. The majority of mechanical systems on earth are composed of rigid or flexible bodies that do not suffer a mass change during their motions. For this reason, Equation 2.2 is most often used in the form,

$$\vec{F} = m\frac{d\vec{v}}{dt} = m\vec{a} \qquad (2.3)$$

where $\vec{a}$ is defined to be the acceleration of the particle.

$$\vec{a} = \frac{d\vec{v}}{dt}. \qquad (2.4)$$

---

1 *Absolute velocity* is defined as the velocity of an object with respect to an inertial reference frame and *absolute acceleration* is defined in the same way. Newton's laws apply only to absolute velocities and accelerations. See Chapter 1.

Both the velocity and the acceleration of the particle relate its motion to an *inertial reference frame* and they are termed the *absolute velocity* and *absolute acceleration* respectively.

2.4 Deriving Equations of Motion for Particles

Newton's laws provide a very convenient method for deriving the equations of motion of simple systems. Equations of motion are the differential equations that, when solved, can be used to predict the response of the system to a set of applied forces.

The procedure for deriving the equations has only four steps and, if they are followed, the derivations are very straightforward. The steps are,

1. *Kinematics* – choose the coordinates (also known as degrees of freedom) to be used to describe the motion and derive expressions for the absolute velocities and accelerations of the masses under consideration. Use the methods of Chapter 1.
2. *Free body diagrams (FBDs)* – sketch the masses under consideration as if they are in space with no forces acting on them. Then add to the sketches all of the externally applied forces acting on the masses. Also add the internal forces of interaction between the masses, being sure that they act in equal and opposite pairs as stipulated by Newton's third law. The FBDs should show the positive sense of the accelerations derived in step 1.
3. *Force balance equations (Newton's second law)* – using the FBDs, equate the vector sum of forces on each body to its mass multiplied by its vector acceleration.
4. *Manipulate and solve the equations* – the first three steps will lead to a set of equations with a set of unknowns. All that is left is to manipulate and combine the equations in order to extract the desired solution.

As an example, we can apply this procedure to the slider in the slot first shown in Figure 1.4 and repeated here for convenience as Figure 2.1. The step by step procedure goes as follows.

Figure 2.1 A slider in a slot.

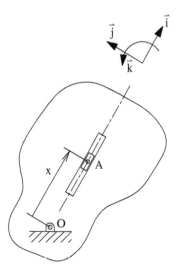

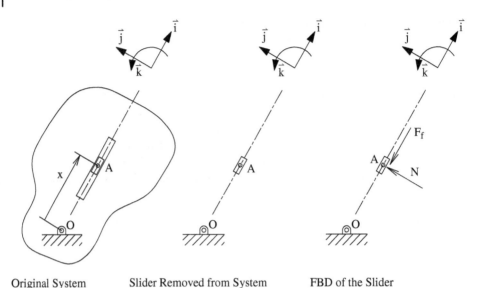

Original System   Slider Removed from System   FBD of the Slider

Figure 2.2   Creating the Free Body Diagram of the slider.

1. *Kinematics* – the absolute velocity and acceleration of the slider have been previously derived in Equations 1.21 and 1.22 and shown to be,

$$\vec{v}_A = \dot{x}\vec{i} + \omega x \vec{j}$$

and,

$$\vec{a}_A = (\ddot{x} - \omega^2 x)\vec{i} + (\dot{\omega}x + 2\omega\dot{x})\vec{j}.$$

Notice that these expressions use the body fixed axes $(\vec{i}, \vec{j}, \vec{k})$ so that the force balances must use the same directions. This is, in fact, a very convenient set of directions for this system since the forces acting on the slider are either aligned with or perpendicular to the slot.

2. *Free body diagrams (FBDs)* – Figure 2.2 shows the FBD of the slider being developed. At first, the slider is simply shown in space, detached from all other bodies and without having any forces acting on it. The forces acting on the slider are then shown along with the positive directions of the accelerations from step 1 which are, in fact, the directions of the unit vectors . Consideration must be given at this stage to the forces that are acting on the slider.

The figure shows two forces.

The first is a normal force N that is acting on the slider due to its interaction with the wall of the slot in the rotating body. The magnitude of this force is not known and its direction is only known to the extent that it must be perpendicular to the line of contact between the slider and the slot. In the figure it is assumed that the force acts in the positive $\vec{j}$ direction on the slider (and, by Newton's third law, in the negative $\vec{j}$ direction on the wall of the slot) although this is not a certainty. The final solution may reveal that N has a negative value and is actually acting in the opposite direction.

The second, F_f, is a frictional force that opposes the motion of the slider along the length of the slot. Frictional forces always oppose the relative motion between two bodies so the force shown in the figure is in a direction that assumes that the slider is moving outward (i.e. the assumption is that $\dot{x}$ is positive). It is not necessary to make this particular assumption but some assumption concerning the direction of action of the force must be made and, once made, must be carried through to the final solution.

We can make the further assumption that the frictional force is due to "dry" or "Coulomb"[2] friction. In its simplest form, this means that the force due to friction can be modeled as being limited to a constant coefficient of friction (usually called μ) multiplied by the normal force between the two bodies in contact. In other words, if the two bodies are assumed not to be moving relative to each other and the force required to prevent them from moving is less than μN, then they will not move and the force will be equal to the required value. If, on the other hand, the two bodies are assumed not to be moving relative to each other and the force required to prevent the motion is greater than μN, then there will be motion and the frictional force will be equal to its maximum value (i.e. $F_f = \mu N$) and will be in a direction opposing the relative motion.

3. *Force balance equations (Newton's second law)* – for this simple system of constant mass (assume the mass of the slider is m), the force balance equations are derived simply by summing the applied forces in the $\vec{i}$ and $\vec{j}$ directions respectively, and equating the total force in each direction to the mass multiplied by the acceleration (derived in step 1) in that direction. The resulting equations are,

$$\sum F_{\vec{i}} : \quad m(\ddot{x} - \omega^2 x) = -F_f \tag{2.5}$$

$$\sum F_{\vec{j}} : \quad m(\dot{\omega} x + 2\omega \dot{x}) = N \tag{2.6}$$

where it should be noted that the positive directions assumed for the forces when drawing the FBD have been maintained.

4. *Manipulate and solve the equations* – given the simplicity of this problem, it is quite remarkable how much information we can get from the equations of motion. There are three cases to consider.

(a) Consider first the case where the slider has been moving in the slot but has now come to rest so that x is a constant. Then, $\dot{x} = \ddot{x} = 0$ and Equations 2.5 and 2.6 simplify to,

$$F_f = m\omega^2 x \tag{2.7}$$

$$N = m\dot{\omega} x. \tag{2.8}$$

The action of specifying that x be constant has introduced a *constraint* on the motion that can only exist if the forces F_f and N are exactly as specified in Equations 2.7 and 2.8. The forces are then known as *constraint forces*.

---

2 Charles-Augustin de Coulomb (1736–1806), a military engineer, won the Grand Prix from the Académie des Sciences of France in 1781 for his work on friction. He also performed groundbreaking work on electricity and magnetism, where he developed the laws of attraction and repulsion, the theory of electric point charges, and the distribution of electricity on the surface of charged bodies.

In fact, the normal force N, in this or any other system, is always a constraint force because it takes on any value required to stop the two bodies from intruding on each other.

Equation 2.8 indicates that the normal force can take on negative values if $\dot{\omega} < 0$. While it is not physically possible for the normal force to hold the two bodies together (only apart), the normal force can become negative by having the slider press against the wall opposite to that assumed in the FBD (Figure 2.2).

Another consideration at this point is whether or not there is sufficient frictional force available for x to remain constant. That is, the maximum available frictional force is $F_{f_{max}} = \pm \mu N$, which, after substituting N from Equation 2.8, is given by $F_{f_{max}} = \pm \mu m \dot{\omega} x$. If the angular acceleration $\dot{\omega}$ should go to zero, for instance, there is no available friction force (i.e. the slider is touching neither of the two walls of the slot) and therefore there is nothing to maintain a constant value of x. If the frictional force required to maintain the constraint is larger in magnitude than the available frictional force, the slider will begin to accelerate along the slot and Equations 2.7 and 2.8 are no longer valid. In that case we must revert to Equations 2.5 and 2.6, which are the governing equations of motion for this system and are always valid.

(b) Consider now the case where the slider has developed a motion outward within the slot. That is, $\dot{x} > 0$ and the frictional force naturally opposes the motion (i.e. acts inward as was assumed when the FBD in Figure 2.2 was constructed). This means that the frictional force is positive and has magnitude,

$$F_f = \mu |N|.$$

Then, substituting for N from Equation 2.6, we find that,

$$F_f = \mu |m(\dot{\omega}x + 2\omega\dot{x}).|$$

This can be substituted into Equation 2.5 to yield the following differential equation governing motion of the slider outward in the slot.

$$\ddot{x} - \omega^2 x + \mu |(\dot{\omega}x + 2\omega\dot{x})| = 0.$$

Except for the absolute value in the equation, it is not a particularly difficult equation to solve. Consider, for example, the case where ω has a positive, known, constant value (say $\omega = \Omega$ and $\dot{\omega} = 0$). The differential equation simplifies to,

$$\ddot{x} + 2\mu\Omega\dot{x} - \Omega^2 x = 0.$$

This is a linear, second order, differential equation with constant coefficients. If we make a term-by-term comparison with the standard equation for a damped oscillator, written in the form,

$$m\ddot{x} + c\dot{x} + kx = 0$$

we can see that the slider has a negative stiffness term that will cause x to grow without bound.

(c) For the final case where the slider has a motion directing it inward within the slot ($\dot{x} < 0$), the frictional force again opposes the motion (i.e. acts outward) and therefore has a negative magnitude in the sense of the force assumed on the FBD. That is,

$$F_f = -\mu |N|.$$

Following a procedure similar to that in the previous case gives the result,

$$\ddot{x} - \omega^2 x - \mu|(\dot{\omega}x + 2\omega\dot{x})| = 0.$$

Considering again the case where ω has a positive, known, constant value (say $\omega = \Omega$ and $\dot{\omega} = 0$). The differential equation simplifies to,

$$\ddot{x} - 2\mu\Omega\dot{x} - \Omega^2 x = 0$$

and the solution again grows without bound, having now both negative stiffness and negative damping.

Part 2 of this text discusses methods of solving the equations of motion.

2.5 Working with Rigid Bodies

To this point, we have considered only the translational motion of particles. We now move on to consider bodies of finite size that comprise groups of particles. These particles are subject to internal constraint forces that cause them to maintain a prescribed geometrical relationship to each other. The simplest constraint causes them to remain a fixed distance from each other and the group of particles becomes a *rigid body*. Another type of constraint will allow the particles to move relative to each other so long as they obey a specified constitutive law. These groups of particles are *flexible bodies*.

Rigid body analysis must take into account the fact that particles within the body have different accelerations since, in the general case, rigid bodies will translate and rotate simultaneously. It is the rotational motion that causes the individual particles to experience different velocities and accelerations.

The concept of *angular momentum* is introduced in Section 2.8 to account for rotational motion. Angular momentum does not arise from a "fourth law of Newton" but is, in fact, derived from what we already know about linear momentum and the relationship of its rate of change to applied forces.

We start the derivation of rigid body equations of motion with a single particle. Figure 2.3 shows a three dimensional rigid body. The body has a general reference point O with known

Figure 2.3 A single particle in a rigid body.

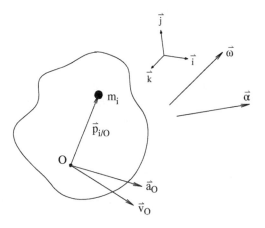

absolute velocity $\vec{v}_O$ and acceleration $\vec{a}_O$. Vectors are expressed in a body-fixed reference frame having unit vectors $(\vec{i}, \vec{j}, \vec{k})$. The body and reference frame have angular velocity $\vec{\omega}$ and angular acceleration $\vec{\alpha}$. The particle to be considered (particle i) is located with respect to the reference point by the position vector $\vec{p}_{i/O}$ and has mass m_i.

2.6 Using $\vec{F} = m\vec{a}$ in the Rigid Body Force Balance

The force balance for the single particle in the rigid body is relatively straightforward. The absolute acceleration of the particle is required and is found to be,

$$\vec{a}_i = \vec{a}_O + \vec{a}_{i/O} \tag{2.9}$$

which, using Equation 1.48 with the understanding that $\vec{p}_{i/O}$ cannot change in length because both points O and i are in the same rigid body[3], becomes,

$$\vec{a}_i = \vec{a}_O + \vec{\alpha} \times \vec{p}_{i/O} + \vec{\omega} \times (\vec{\omega} \times \vec{p}_{i/O}). \tag{2.10}$$

The FBD of the particle is presented in Figure 2.4 where two forces are shown being applied to the particle. The first of these, $\vec{F}_i$, represents the vector sum of all externally applied forces acting on the particle. This is a force arising from the interaction of the rigid body with its external environment. The second applied force, $\vec{f}_i$, is the vector sum of internal forces of constraint acting on particle i. This is the force of interaction between particle i and its neighboring particles that enforces the rigid body constraint. Forces of this type must occur in equal and opposite pairs on neighboring particles in accord with Newton's third law.

According to Newton's second law, and assuming that mass does not change with time, the vector sum of the two forces acting on the particle must be equal to its mass multiplied by its acceleration, as follows,

$$\vec{f}_i + \vec{F}_i = m_i \vec{a}_i \tag{2.11}$$

where $\vec{a}_i$ is known from Equation 2.10 and can be substituted into Equation 2.11 to yield,

$$\vec{f}_i + \vec{F}_i = m_i[\vec{a}_O + \vec{\alpha} \times \vec{p}_{i/O} + \vec{\omega} \times (\vec{\omega} \times \vec{p}_{i/O})] \tag{2.12}$$

or,

$$\vec{f}_i + \vec{F}_i = m_i \vec{a}_O + m_i(\vec{\alpha} \times \vec{p}_{i/O}) + m_i[\vec{\omega} \times (\vec{\omega} \times \vec{p}_{i/O})]. \tag{2.13}$$

Consider the case where Equation 2.13 has been written for every particle in the rigid body. We can then add together all such equations to get,

$$\sum_i \vec{f}_i + \sum_i \vec{F}_i = \sum_i m_i \vec{a}_O + \sum_i m_i(\vec{\alpha} \times \vec{p}_{i/O}) + \sum_i m_i[\vec{\omega} \times (\vec{\omega} \times \vec{p}_{i/O})] \tag{2.14}$$

---

3 The *rigid body constraint* dictates that any two points in the same rigid body must maintain a constant separation distance. This is, in fact, the definition of a rigid body. Points in *elastic bodies* are not subject to this constraint. In reality every body has some degree of elasticity but the assumption of rigid bodies can often be made without loss of accuracy.

Figure 2.4 Free body diagram of a single particle in a rigid body.

where $\sum\limits_{i}$ indicates that the summation is over all particles i in the body. Noting that m_i is a scalar, it is possible to move it around within Equation 2.14 without affecting the result[4]. Terms that do not depend on the index of summation i may also be moved outside the summation process. As a result, Equation 2.14 can be rewritten as,

$$\sum_{i} \vec{f_i} + \sum_{i} \vec{F_i} = \left(\sum_{i} m_i\right) \vec{a}_O + \vec{\alpha} \times \left(\sum_{i} m_i \vec{p}_{i/O}\right) + \vec{\omega} \times \left[\vec{\omega} \times \left(\sum_{i} m_i \vec{p}_{i/O}\right)\right].$$

(2.15)

There are several terms of interest in Equation 2.15. These are,

1. $\sum\limits_{i} \vec{f_i}$ – this is the sum, over all the particles in the rigid body of the internal forces of interaction. By Newton's third law, all of these forces appear in equal and opposite pairs and the vector sum of them over the body must be zero since each pair adds to zero. The result is,

$$\sum_{i} \vec{f_i} = 0.$$

2. $\sum\limits_{i} \vec{F_i}$ – this is the vector sum of all externally applied forces acting on the body. It includes forces that act on all particles (e.g. gravity forces or magnetic forces) and those that act only on one or a few particles (e.g. a force applied at a point). We write the total externally applied force as $\vec{F}$ and make the substitution,

$$\sum_{i} \vec{F_i} = \vec{F}.$$

3. $\sum\limits_{i} m_i$ – this is clearly the sum of the masses of all particles making up the rigid body and, as a result, is the total mass of the body. Let the total mass of the body be called M, so that,

$$\sum_{i} m_i = M.$$

4. $\sum\limits_{i} m_i \vec{p}_{i/O}$ – this term is a little more difficult to explain than the others and appears twice. Each component of the summation is the product of a mass (i.e. m_i) and a position (both distance and direction) relative to the reference point O (i.e. $\vec{p}_{i/O}$). The sum of all such terms will therefore be the product of a mass and a position vector.

---

4 It is often not possible to change the order of vector operations without affecting the result so care should be taken if tempted to do so.

In fact, the sum of all of the individual terms of this type will be equal to the total mass of the body multiplied by the position of the *center of mass* of the body relative to the reference point O[5]. The center of mass is commonly designated as the point G and the position of G relative to O can be written as $\vec{p}_{G/O}$, yielding,

$$\sum_i m_i \vec{p}_{i/O} = M \vec{p}_{G/O}.$$

Substituting these definitions into Equation 2.15 and moving the scalar mass M outside the vector operations simplifies it to,

$$\vec{F} = M\vec{a}_O + M(\vec{\alpha} \times \vec{p}_{G/O}) + M[\vec{\omega} \times (\vec{\omega} \times \vec{p}_{G/O})] \tag{2.16}$$

or, even more simply,

$$\vec{F} = M[\vec{a}_O + \vec{\alpha} \times \vec{p}_{G/O} + \vec{\omega} \times (\vec{\omega} \times \vec{p}_{G/O})] \tag{2.17}$$

where the expression,

$$\vec{a}_O + \vec{\alpha} \times \vec{p}_{G/O} + \vec{\omega} \times (\vec{\omega} \times \vec{p}_{G/O})$$

is easily recognized as the absolute acceleration of the center of mass G. That is, applying Equation 2.10 to find the acceleration of G rather than i would give,

$$\vec{a}_G = \vec{a}_O + \vec{\alpha} \times \vec{p}_{G/O} + \vec{\omega} \times (\vec{\omega} \times \vec{p}_{G/O}) \tag{2.18}$$

which is exactly the same result.

The force balance equation for the rigid body is therefore completely analogous to that used for a particle. That is,

$$\vec{F} = M\vec{a}_G.$$

2.7 Using $\vec{F} = \frac{d\vec{G}}{dt}$ in the Rigid Body Force Balance

Section 2.6 presented a derivation of the force balance relationship for a rigid body using Newton's second law in the form $\vec{F} = m\vec{a}$. Here, we treat the case again using the more general form of the second law, $\vec{F} = \frac{d\vec{G}}{dt}$.

Using Equation 2.1, the force balance equation for the particle i shown in Figure 2.4 can be written as,

$$\vec{f}_i + \vec{F}_i = \frac{d\vec{G}_i}{dt}. \tag{2.19}$$

Equation 2.19 can be written for every particle in the body and all of the resulting equations can be added together to give,

$$\sum_i \vec{f}_i + \sum_i \vec{F}_i = \sum_i \frac{d\vec{G}_i}{dt} = \frac{d}{dt}\left(\sum_i \vec{G}_i\right). \tag{2.20}$$

---

5 Readers are encouraged to review the process used in determining the location of the center of mass of a body. Most reference books will discuss this process in one dimension only but three successive applications of their procedures in three orthogonal directions can readily be visualized as the vector approach used here.

The linear momentum of particle i in Figure 2.3 is simply the mass m_i multiplied by the absolute velocity $\vec{v}_i$, where,

$$\vec{v}_i = \vec{v}_O + \vec{\omega} \times \vec{p}_{i/O}.$$

That is,

$$\vec{G}_i = m_i(\vec{v}_O + \vec{\omega} \times \vec{p}_{i/O}) = (m_i)\vec{v}_O + \vec{\omega} \times (m_i \vec{p}_{i/O}) \tag{2.21}$$

and, the total linear momentum of the body is given by,

$$\vec{G} = \sum_i \vec{G}_i = \left(\sum_i m_i \right) \vec{v}_O + \vec{\omega} \times \left(\sum_i m_i \vec{p}_{i/O} \right). \tag{2.22}$$

As explained in Section 2.6, the expression $\sum_i m_i$ is equal to the total mass of the body M and the expression $\sum_i m_i \vec{p}_{i/O}$ is equal to the total mass of the body multiplied by the position of the center of mass relative to the reference point O (i.e. $M\vec{p}_{G/O}$).

As a result, the total linear momentum of the body can be written as,

$$\vec{G} = M(\vec{v}_O + \vec{\omega} \times \vec{p}_{G/O}). \tag{2.23}$$

Clearly, $\vec{v}_O + \vec{\omega} \times \vec{p}_{G/O}$ is the expression for the velocity of the center of mass of the body (i.e. $\vec{v}_G$) and the total linear momentum is, in fact,

$$\vec{G} = M\vec{v}_G. \tag{2.24}$$

Then, combining Equations 2.22 and 2.24, shows that,

$$\sum_i \vec{G}_i = M\vec{v}_G. \tag{2.25}$$

This can be substituted into Equation 2.20 to give,

$$\sum_i \vec{f}_i + \sum_i \vec{F}_i = \frac{d}{dt}(M\vec{v}_G) = \frac{d\vec{G}}{dt}. \tag{2.26}$$

As in Section 2.6,

$$\sum_i \vec{f}_i = 0 \tag{2.27}$$

and

$$\sum_i \vec{F}_i = F \tag{2.28}$$

where F is the total applied force. As a result, the force balance equation for the body can be written as,

$$\vec{F} = \frac{d\vec{G}}{dt}. \tag{2.29}$$

A force balance for a rigid body therefore can be handled in one of two ways, both of which require that the location of the center of mass, G, be known,

1. Using the absolute acceleration of the center of mass,

$$\vec{F} = M\vec{a}_G.$$

2. Using the linear momentum of the rigid body,

$$\vec{F} = \frac{\mathrm{d}\vec{G}}{\mathrm{d}t} = \frac{\mathrm{d}}{\mathrm{d}t}(M\vec{v}_G).$$

2.8 Moment Balance for a Rigid Body

Performing a moment balance on a rigid body requires an understanding of the process of "taking moments" about a point. We will use the reference point O defined in Section 2.6 and, particularly, in Figure 2.3.

If a force acts on a point in the body, then the moment or torque produced by that force around point O is defined to be,

$$\vec{M}_O = \vec{p}_{F/O} \times \vec{F} \tag{2.30}$$

where $\vec{M}_O$ is the moment about O, $\vec{F}$ is the applied force, and $\vec{p}_{F/O}$ is the position vector from O to the point of application of the applied force[6].

Consider the total moment about point O resulting from the two forces applied on the particle i. Remember that the forces are an externally applied force, $\vec{F}_i$, and an internal constraint force, $\vec{f}_i$. The moment about O due to these forces is therefore,

$$\vec{M}_O = \vec{p}_{i/O} \times \vec{f}_i + \vec{p}_{i/O} \times \vec{F}_i = \vec{p}_{i/O} \times (\vec{f}_i + \vec{F}_i). \tag{2.31}$$

Using Equation 2.19 with Equation 2.31, the following equality can be written.

$$\vec{p}_{i/O} \times (\vec{f}_i + \vec{F}_i) = \vec{p}_{i/O} \times \frac{\mathrm{d}\vec{G}_i}{\mathrm{d}t}. \tag{2.32}$$

As was done for the force balance, we consider the case where Equation 2.32 has been written for every particle in the body and the resulting equations have all been added together to give,

$$\sum_i \vec{p}_{i/O} \times \vec{f}_i + \sum_i \vec{p}_{i/O} \times \vec{F}_i = \sum_i \vec{p}_{i/O} \times \frac{\mathrm{d}\vec{G}_i}{\mathrm{d}t}. \tag{2.33}$$

Two of the terms in Equation 2.33 can be dealt with immediately. These are,

1. $\sum_i \vec{p}_{i/O} \times \vec{f}_i$: this is the total moment about the reference point O caused by internal forces. Since these forces occur in equal and opposite pairs, the moment caused by any given force is canceled out by its counterpart and, $\sum_i \vec{p}_{i/O} \times \vec{f}_i = 0$.

---

6 Most engineers are more familiar with 2D moments where you simply multiply a force by a distance and visualize the direction of the moment. This vector approach includes both the magnitude and the direction of the moment and is indispensable in 3D problems.

2. $\sum_i \vec{p}_{i/O} \times \vec{F}_i$: this is the total moment about the reference point O caused by externally applied forces. Let this be denoted by $\vec{M}_O$.

Equation 2.33 can then be rewritten as,

$$\vec{M}_O = \sum_i \vec{p}_{i/O} \times \frac{d\vec{G}_i}{dt}. \tag{2.34}$$

The third term (i.e. $\sum_i \vec{p}_{i/O} \times \frac{d\vec{G}_i}{dt}$) is more difficult to deal with.

We start by defining the *angular momentum* of the rigid body about point O. *In the same way that we defined moments caused by forces as a vector cross product, we define the angular momentum about point O due to the particle i (i.e. $\vec{H}_{i/O}$) to be the moment of the linear momentum of i about O.* That is,

$$\vec{H}_{i/O} = \vec{p}_{i/O} \times \vec{G}_i. \tag{2.35}$$

It is then possible to express the total angular momentum of the rigid body about the reference point as,

$$\vec{H}_O = \sum_i \vec{p}_{i/O} \times \vec{G}_i. \tag{2.36}$$

We differentiate $\vec{H}_O$ with respect to time and find the third term of Equation 2.33,

$$\frac{d\vec{H}_O}{dt} = \sum_i \frac{d\vec{p}_{i/O}}{dt} \times \vec{G}_i + \underbrace{\sum_i \vec{p}_{i/O} \times \frac{d\vec{G}_i}{dt}}_{\text{the third term!}}. \tag{2.37}$$

We reorder this to write the third term as,

$$\sum_i \vec{p}_{i/O} \times \frac{d\vec{G}_i}{dt} = \frac{d\vec{H}_O}{dt} - \sum_i \frac{d\vec{p}_{i/O}}{dt} \times \vec{G}_i. \tag{2.38}$$

Substituting $m_i \vec{v}_i$ for $\vec{G}_i$ gives,

$$\sum_i \vec{p}_{i/O} \times \frac{d\vec{G}_i}{dt} = \frac{d\vec{H}_O}{dt} - \sum_i \frac{d\vec{p}_{i/O}}{dt} \times (m_i \vec{v}_i). \tag{2.39}$$

At this point, we can substitute Equation 2.39 into Equation 2.34 to get,

$$\vec{M}_O = \frac{d\vec{H}_O}{dt} - \sum_i \frac{d\vec{p}_{i/O}}{dt} \times (m_i \vec{v}_i). \tag{2.40}$$

Notice that $\frac{d\vec{p}_{i/O}}{dt}$ is simply an expression for the velocity of i with respect to O (i.e. $\frac{d\vec{p}_{i/O}}{dt} = \vec{v}_{i/O}$) and we can write,

$$\vec{M}_O = \frac{d\vec{H}_O}{dt} - \sum_i \vec{v}_{i/O} \times (m_i \vec{v}_i) = \frac{d\vec{H}_O}{dt} - \sum_i m_i (\vec{v}_{i/O} \times \vec{v}_i). \tag{2.41}$$

The absolute velocity of i, $\vec{v}_i$, can be written as $\vec{v}_i = \vec{v}_O + \vec{v}_{i/O}$. Substituting this expression in Equation 2.41 and expanding yields,

$$\vec{M}_O = \frac{d\vec{H}_O}{dt} - \sum_i (m_i \vec{v}_{i/O}) \times \vec{v}_O - \sum_i m_i (\vec{v}_{i/O} \times \vec{v}_{i/O}).$$ (2.42)

Since $\vec{v}_{i/O} \times \vec{v}_{i/O} = 0$ by definition, and

$$\sum_i m_i \vec{v}_{i/O} = \sum_i m_i \frac{d\vec{p}_{i/O}}{dt} = \frac{d}{dt}\left(\sum_i m_i \vec{p}_{i/O} \right).$$

Equation 2.42 can be rewritten as,

$$\vec{M}_O = \frac{d\vec{H}_O}{dt} - \left[\frac{d}{dt}\left(\sum_i m_i \vec{p}_{i/O} \right) \right] \times \vec{v}_O.$$ (2.43)

The term including the mass of particle i was previously discussed in Section 2.6 (see the material following Equation 2.15) and found to be equal to the total mass of the body M multiplied by the position of the center of mass with respect to the reference point O, $\vec{p}_{G/O}$. Making this substitution, and noting that the mass of a rigid body is constant, gives,

$$\vec{M}_O = \frac{d\vec{H}_O}{dt} - \left[\frac{d}{dt}(M\vec{p}_{G/O}) \right] \times \vec{v}_O = \frac{d\vec{H}_O}{dt} - M(\vec{v}_{G/O} \times \vec{v}_O).$$ (2.44)

We now note that the velocity of G with respect to O can be rewritten as $\vec{v}_{G/O} = \vec{v}_G - \vec{v}_O$ simply by rearranging the relative velocity expression $\vec{v}_G = \vec{v}_O + \vec{v}_{G/O}$. Substituting this into Equation 2.44 and noting that $\vec{v}_O \times \vec{v}_O = 0$ gives,

$$\vec{M}_O = \frac{d\vec{H}_O}{dt} - M(\vec{v}_G \times \vec{v}_O).$$ (2.45)

Equation 2.45 is the general form of the moment balance for a rigid body where the moments are taken about an arbitrary, moving reference point O. There are two special reference points that are more often used for the moment balance. These are,

1. The case where the reference point is the center of mass. That is, O and G are coincident. In this case, the term $M(\vec{v}_G \times \vec{v}_O)$ is equal to zero because $\vec{v}_O = \vec{v}_G$ and the moment balance equation becomes,

$$\vec{M}_G = \frac{d\vec{H}_G}{dt}.$$

2. The case where the reference point has zero velocity. That is, point O is a "fixed point" by virtue of being connected to ground through a joint. In this case, $\vec{v}_O = 0$ and the moment balance is,

$$\vec{M}_O = \frac{d\vec{H}_O}{dt}.$$

A moment balance for a rigid body therefore can be handled in three possible ways,

1. Using a moving reference point O on the rigid body,

$$\vec{M}_O = \frac{d\vec{H}_O}{dt} - M(\vec{v}_G \times \vec{v}_O).$$

2. Using the center of mass of the rigid body as the reference point,

$$\vec{M}_G = \frac{d\vec{H}_G}{dt}.$$

3. Using fixed point O on the rigid body as the reference point,

$$\vec{M}_O = \frac{d\vec{H}_O}{dt}.$$

2.9 The Angular Momentum Vector – $\vec{H}_O$

Before being able to put the moment balance equations from Section 2.8 to use, we must first consider generating the angular momentum vector for the rigid body.

The angular momentum vector, $\vec{H}$, cannot be defined without reference to some point on the rigid body in question[7]. This is because *angular momentum is defined as the moment of linear momentum*. Equation 2.35 defines angular momentum about the reference point O due to the mass particle i as,

$$\vec{H}_{i/O} = \vec{p}_{i/O} \times \vec{G}_i \tag{2.46}$$

where the linear momentum is defined to be $\vec{G}_i = m_i \vec{v}_i$, resulting in,

$$\vec{H}_{i/O} = m_i (\vec{p}_{i/O} \times \vec{v}_i). \tag{2.47}$$

We assume that the absolute velocity of the reference point, $\vec{v}_O$, and the angular velocity of the rigid body, $\vec{\omega}$, are known. We can then write an expression for the velocity of particle i (noting that the position vector $\vec{p}_{i/O}$ has no rate of change of magnitude because both i and O are in the same rigid body and cannot move apart) as,

$$\begin{aligned}
\vec{v}_i &= \vec{v}_O + \vec{v}_{i/O} \\
&= \vec{v}_O + \frac{d}{dt}(\vec{p}_{i/O}) \\
&= \vec{v}_O + \vec{\omega} \times \vec{p}_{i/O}. \tag{2.48}
\end{aligned}$$

Referring to Figure 2.3, we can write expressions for the position of i with respect to O and the absolute velocity of point O in the reference frame with unit vectors $(\vec{i}, \vec{j}, \vec{k})$.

Let the distance in the $\vec{i}$ direction from point O to point i be x_i. Similarly, we define y_i and z_i to be the distances in the $\vec{j}$ and $\vec{k}$ directions respectively. The position vector is then,

$$\vec{p}_{i/O} = x_i \vec{i} + y_i \vec{j} + z_i \vec{k}.$$

Let the absolute velocity of O have scalar components v_{Ox}, v_{Oy}, and v_{Oz}. The velocity of O is then,

$$\vec{v}_O = v_{Ox} \vec{i} + v_{Oy} \vec{j} + v_{Oz} \vec{k}.$$

---

7 There are instances during the analysis of dynamic systems where it may be advantageous to define the angular momentum about a point that is not on the body being analyzed. Often this is done for bodies which can be approximated as particles. For rigid bodies, the reference point is always located on the body.

Further, let the angular velocity of the coordinate system $(\vec{i}, \vec{j}, \vec{k})$ be,

$$\vec{\omega} = \omega_x \vec{i} + \omega_y \vec{j} + \omega_z \vec{k}$$

The cross product $\vec{\omega} \times \vec{p}_{i/O}$ can then be written as,

$$\vec{\omega} \times \vec{p}_{i/O} = [\omega_y z_i - \omega_z y_i]\vec{i} + [\omega_z x_i - \omega_x z_i]\vec{j} + [\omega_x y_i - \omega_y x_i]\vec{k}$$

and the velocity of particle i is,

$$\vec{v}_i = [v_{Ox} + \omega_y z_i - \omega_z y_i]\vec{i} + [v_{Oy} + \omega_z x_i - \omega_x z_i]\vec{j} + [v_{Oz} + \omega_x y_i - \omega_y x_i]\vec{k}. \tag{2.49}$$

Equation 2.49 can be substituted into Equation 2.47 to give, after performing another cross product and gathering some terms, an expression for the angular momentum about O due to particle i as follows:

$$\begin{aligned}
\vec{H}_{i/O} = {} & m_i[y_i v_{Oz} - z_i v_{Oy} + (y_i^2 + z_i^2)\omega_x - x_i y_i \omega_y - x_i z_i \omega_z]\vec{i} \\
& + m_i[z_i v_{Ox} - x_i v_{Oz} - y_i x_i \omega_x + (x_i^2 + z_i^2)\omega_y - y_i z_i \omega_z]\vec{j} \\
& + m_i[x_i v_{Oy} - y_i v_{Ox} - z_i x_i \omega_x - z_i y_i \omega_y + (x_i^2 + y_i^2)\omega_z]\vec{k}. \tag{2.50}
\end{aligned}$$

To get the total angular momentum vector about point O, we write Equation 2.50 for every particle and add the resulting equations together to get,

$$\vec{H}_O = \sum_i \vec{H}_{i/O} \tag{2.51}$$

or,

$$\begin{aligned}
\vec{H}_O = {} & [M\bar{y}v_{Oz} - M\bar{z}v_{Oy} + I_{xx_O}\omega_x - I_{xy_O}\omega_y - I_{xz_O}\omega_z]\vec{i} \\
& + [M\bar{z}v_{Ox} - M\bar{x}v_{Oz} - I_{yx_O}\omega_x + I_{yy_O}\omega_y - I_{yz_O}\omega_z]\vec{j} \\
& + [M\bar{x}v_{Oy} - M\bar{y}v_{Ox} - I_{zx_O}\omega_x - I_{zy_O}\omega_y + I_{zz_O}\omega_z]\vec{k}
\end{aligned}$$

$$\tag{2.52}$$

where the terms used in Equation 2.52 are defined as follows.

1. $M = \sum_i m_i$. This is the total mass of the rigid body.

2. $M\bar{x} = \sum_i (m_i x_i)$. This summation is one component of that used to locate the center of mass of the body with respect to the reference point. $\bar{x}$ is the x-component of $\vec{p}_{G/O}$.

3. $M\bar{y} = \sum_i (m_i y_i)$. $\bar{y}$ is the y-component of $\vec{p}_{G/O}$.

4. $M\bar{z} = \sum_i (m_i z_i)$. $\bar{z}$ is the z-component of $\vec{p}_{G/O}$.

5. $I_{xx_O} = \sum_i m_i(y_i^2 + z_i^2)$. This term is a function of the spatial distribution of mass particles around the x-axis (i.e. $\vec{i}$ direction) passing through point O. The term is always positive because of the sum of squares term. It is called the *x moment of inertia of the body about point O*.

6. $I_{xy_O} = \sum_i m_i x_i y_i$. This term is also a function of the distribution of the mass in the body but it has the potential to be negative or positive depending upon the signs of x_i and y_i. It is called the *x–y product of inertia about point O*.

7. $I_{xz_O} = \sum_i m_i x_i z_i$. This is the x–z product of inertia about point O.

8. $I_{yx_O} = \sum_i m_i y_i x_i$. This is the y–x product of inertia about point O. Note that $I_{yx_O} = I_{xy_O}$

   since the order of the x and y terms in the summations do not change the result.

9. $I_{yy_O} = \sum_i m_i(x_i^2 + z_i^2)$. This is the y moment of inertia of the body about point O.

10. $I_{yz_O} = \sum_i m_i y_i z_i$. The y–z product of inertia about O.

11. $I_{zx_O} = \sum_i m_i z_i x_i$. The z–x product of inertia about O. Note that this is equal to the x–z

    product of inertia.

12. $I_{zy_O} = \sum_i m_i z_i y_i$. The z–y product of inertia about O, which is equal to the y–z product

    of inertia about O.

13. $I_{zz_O} = \sum_i m_i(x_i^2 + y_i^2)$. The z moment of inertia about point O.

In fact, the particle of mass we have been considering, m_i, can be considered to be infinitesimally small. In this case the summations over all particles in the rigid body can be replaced by integrations. The mass particle becomes dm and the moments and products of inertia are,

$$I_{xx_O} = \int_{body} (y^2 + z^2)\, dm \tag{2.53}$$

$$I_{yy_O} = \int_{body} (x^2 + z^2)\, dm \tag{2.54}$$

$$I_{zz_O} = \int_{body} (x^2 + y^2)\, dm \tag{2.55}$$

$$I_{xy_O} = I_{yx_O} = \int_{body} xy\, dm \tag{2.56}$$

$$I_{xz_O} = I_{zx_O} = \int_{body} xz\, dm \tag{2.57}$$

$$I_{yz_O} = I_{zy_O} = \int_{body} yz\, dm. \tag{2.58}$$

Referring to Equation 2.52 and remembering that we wish to restrict the moment balance on the body to be about a reference point, which is either the center of mass ($\bar{x} = \bar{y} = \bar{z} = 0$) or a "fixed point" ($v_{Ox} = v_{Oy} = v_{Oz} = 0$), we can now write the angular momentum vector for the body as,

$$\vec{H}_O = [I_{xx_O}\omega_x - I_{xy_O}\omega_y - I_{xz_O}\omega_z]\vec{i}$$
$$+ [-I_{yx_O}\omega_x + I_{yy_O}\omega_y - I_{yz_O}\omega_z]\vec{j}$$
$$+ [-I_{zx_O}\omega_x - I_{zy_O}\omega_y + I_{zz_O}\omega_z]\vec{k} \tag{2.59}$$

or, if we write the vector as,

$$\vec{H}_O = H_x\vec{i} + H_y\vec{j} + H_z\vec{k} \tag{2.60}$$

the components of $\vec{H}_O$ can be found from,

$$\begin{Bmatrix} H_x \\ H_y \\ H_z \end{Bmatrix}_O = \begin{bmatrix} I_{xx_O} & -I_{xy_O} & -I_{xz_O} \\ -I_{yx_O} & I_{yy_O} & -I_{yz_O} \\ -I_{zx_O} & -I_{zy_O} & I_{zz_O} \end{bmatrix} \begin{Bmatrix} \omega_x \\ \omega_y \\ \omega_z \end{Bmatrix}. \tag{2.61}$$

We can write Equation 2.61 in general as,

$$\{H\}_O = [I_O]\{\omega\} \tag{2.62}$$

where $[I_O]$ is known as the *inertia tensor*.

For the moment balance we have been considering, we need the angular momentum vector only to write its derivative with respect to time. As with all vectors, the angular momentum vector will have a rate of change of magnitude and a rate of change of direction. Of these, the rate of change of direction is by far the most interesting because it leads to *gyroscopic effects*, which will be discussed more fully later. The derivative of the angular momentum vector is,

$$\frac{d\vec{H}_O}{dt} = \underbrace{\dot{H}_x\,\vec{i} + \dot{H}_y\,\vec{j} + \dot{H}_z\,\vec{k}}_{\substack{\text{rate of change} \\ \text{of magnitude}}} + \underbrace{\vec{\omega} \times \vec{H}_O}_{\substack{\text{rate of change} \\ \text{of direction}}} . \tag{2.63}$$

2.10 A Physical Interpretation of Moments and Products of Inertia

Consider the simple system shown in Figure 2.5. There is a single mass particle m that is constrained by a massless rigid link to remain a fixed distance from the point O. We consider the case where the mass moves in a circular horizontal path in three dimensional space. A coordinate system $(\vec{i}, \vec{j}, \vec{k})$ is defined so that $\vec{k}$ is aligned with the axis of rotation of the mass, and $\vec{i}$ and $\vec{j}$ rotate about $\vec{k}$ so that the mass remains in the plane formed by $\vec{i}$ and $\vec{k}$. The position of the mass with respect to point O is therefore,

$$\vec{p}_{m/O} = x\vec{i} + z\vec{k} \tag{2.64}$$

Figure 2.5 A single mass system to help interpret inertia properties.

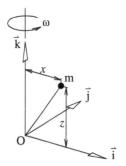

where the distances x and z are constant with time.

Let the angular velocity of the reference frame be,

$$\vec{\omega} = \omega \vec{k} \tag{2.65}$$

where the magnitude ω is not constant. The mass therefore has an angular acceleration of,

$$\vec{\alpha} = \dot{\omega}\vec{k}. \tag{2.66}$$

The reason for choosing such a simple system is that the expression for the angular momentum is relatively simple but still contains all of the terms needed to see how moments and products of inertia enter the equations of motion. Considering the expression for the angular momentum vector given in Equation 2.59 and setting $\omega_z = \omega$ as well as $\omega_x = \omega_y = 0$ results in,

$$\vec{H}_O = (-I_{xz_0}\omega)\vec{i} + (-I_{yz_0}\omega)\vec{j} + (I_{zz_0}\omega)\vec{k}. \tag{2.67}$$

The integrations over the body required to find the moment of inertia (I_{zz_0}) and the products of inertia (I_{xz_0} and I_{yz_0}) are made simple by the fact that there is only a single mass particle and that its $\vec{j}$ coordinate is zero (i.e. $y = 0$). The resulting expressions are,

$$I_{xz_0} = \int_{body} xz\,dm = m \times x \times z = mxz$$

$$I_{yz_0} = \int_{body} yz\,dm = m \times 0 \times z = 0$$

$$I_{zz_0} = \int_{body} (x^2 + y^2)\,dm = m \times (x^2 + 0^2) = mx^2. \tag{2.68}$$

Making these substitutions in Equation 2.67 results in a simplified expression for the angular momentum vector, as follows.

$$\vec{H}_O = (-mxz\omega)\vec{i} + (mx^2\omega)\vec{k}. \tag{2.69}$$

Applying the expression for the moment balance about a fixed point

$$\vec{M}_O = \frac{d\vec{H}_O}{dt}$$

to Equation 2.69 (noting that, since x and z are constant, $\dot{x}$ and $\dot{z}$ are both zero) yields,

$$\vec{M}_O = (-mxz\dot{\omega})\vec{i} + (mx^2\dot{\omega})\vec{k} + \omega\vec{k} \times \vec{H}_O \tag{2.70}$$

or, after performing the cross product,

$$\vec{M}_O = \underbrace{(-mxz\dot{\omega})\vec{i}}_{1} + \underbrace{(-mxz\omega^2)\vec{j}}_{2} + \underbrace{(mx^2\dot{\omega})\vec{k}}_{3}. \tag{2.71}$$

Before considering each of the terms in Equation 2.71 it is necessary to have an expression for the acceleration of the mass. As usual, the derivation of an expression for the acceleration starts with the position vector and differentiates twice, as follows, again noting that $\dot{x}$

and $\dot{y}$ are both zero,

$$\vec{p}_{m/O} = x\vec{i} + z\vec{k}$$

$$\vec{v}_m = \frac{d\vec{p}_{m/O}}{dt} = (\dot{x}\vec{i} + \dot{z}\vec{k}) + (\omega\vec{k}) \times (x\vec{i} + z\vec{k}) = \omega x\vec{j}$$

$$\vec{a}_m = \frac{d\vec{v}_m}{dt} = (\dot{\omega}x + \omega\dot{x})\vec{j} + (\omega\vec{k}) \times (\omega x\vec{j}) = -\omega^2 x\vec{i} + \dot{\omega}x\vec{j}. \tag{2.72}$$

The terms in Equation 2.71 can be explained as follows.

1. The first term is a product of inertia term. This can be determined from the form of the term where two different coordinates are multiplied together rather than having coordinates squared as would be expected in a moment of inertia term.

   The term is, in fact, an expression of the moment which must be applied about the x-axis in order to have the specified motion of the mass. Note that the problem being considered is not one where forces are applied and the equations of motion are used to predict the resulting motion. Instead, the motion has been predetermined (i.e. circular motion in a horizontal plane) and the forces that cause the motion can be found from Newton's second law. In other words, knowing the acceleration of the mass (see Equation 2.72), the force applied to the mass is inferred to be,

   $$\vec{F}_m = -m\omega^2 x\vec{i} + m\dot{\omega}x\vec{j}. \tag{2.73}$$

   Figure 2.6 shows the mass as viewed from a point on the positive x-axis. From this point, we can see the $\vec{j}$ component of $\vec{F}_m$ and its moment arm z. The resulting component of the moment about O is $M_{O_1} = -mxz\dot{\omega}$. The negative sign indicates that the moment is in a direction opposite to a rotation about the positive $\vec{i}$ direction. This is exactly equal to "term 1" from Equation 2.71 and indicates that this product of inertia term accounts for the moments required to maintain the tangential acceleration $x\dot{\omega}$. While we have considered only a single particle here, this statement would be true for every particle in a general rigid body and the summation of all such terms leads to the integral over the body which ultimately leads to the total x–z product of inertia of the body.

   The part of the figure shown in dashed lines indicates what would happen if there were another mass symmetrically placed in the body. This mass produces a moment that

Figure 2.6   Term 1 – a product of inertia term.

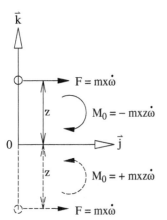

Figure 2.7 Term 2 – a product of inertia term.

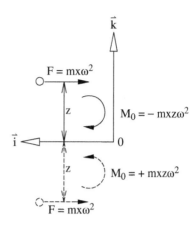

is equal and opposite to that produced by the mass we have been considering. If both masses were present, there would be a net moment of zero about point O in this plane. This illustrates that products of inertia for objects that are symmetric in a plane are zero.

2. The second term is also a product of inertia term. In this case, it is the moment which must be applied about the y-axis in order for the mass to maintain its rotation in the x–y plane.

   The centripetal acceleration component, $x\omega^2$, must exist during the circular motion and there must therefore be a force equal to $mx\omega^2$ acting on the mass as shown in Figure 2.7 (a view from a point on the positive y-axis). This force exerts a negative moment, relative to the y-axis, about point O. The magnitude of the moment depends upon the magnitude of the force and the length of the moment arm, in this case z. The moment is therefore $M_O = -mxz\omega^2$. This is exactly "term 2". Once again, dashed lines in Figure 2.7 indicate what would happen if another, symmetrically placed, mass was present. Since this is a product of inertia term, the two moments cancel each other.

3. The third term is different from the others because it displays the squared coordinate (in this case x^2) characteristic of a moment of inertia. Figure 2.8 shows a view from a point

Figure 2.8 Term 3 – a moment of inertia term.

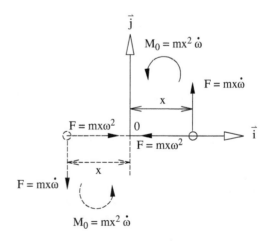

on the positive z-axis and we can see the same force that we saw in Figure 2.6. That is, the force necessary to produce the tangential acceleration $x\dot{\omega}$. The difference in this view is that the moment arm is now equal to the radius in the acceleration term. The result is that the moment is $M_O = mx^2\dot{\omega}$.

The force causing the centripetal acceleration $(F = mx\omega^2)$ can also be seen on Figure 2.8 but, since the force acts through the reference point O, it does not contribute to the moment.

The dashed lines indicating the effect of a symmetrically placed mass in Figure 2.8 show a moment that adds to that already there. Moments of inertia are always additive. Extra mass simply makes the moment of inertia larger. As a result, moments of inertia are never negative and are only zero when considering a particle or a line of particles on the axis of rotation. If mass is distributed away from the axis of rotation, there will be a non-zero moment of inertia.

2.11   Euler's Moment Equations

Leonhard Euler[8] developed a form of the moment balance equations for a rigid body.

Given that the angular momentum vector $\vec{H}_O$ can be expressed as three scalar components in a reference frame $(\vec{i}, \vec{j}, \vec{k})$ fixed in the rigid body,

$$\vec{H}_O = H_x\vec{i} + H_y\vec{j} + H_z\vec{k}$$

and that the rigid body has angular velocity,

$$\vec{\omega} = \omega_x\vec{i} + \omega_y\vec{j} + \omega_z\vec{k}$$

then,

$$\frac{d\vec{H}_O}{dt} = \dot{\vec{H}}_O + \vec{\omega} \times \vec{H}_O. \tag{2.74}$$

Expanding Equation 2.74 yields,

$$\begin{aligned}
\frac{d\vec{H}_O}{dt} &= (\dot{H}_x - H_y\omega_z + H_z\omega_y)\vec{i} \\
&\quad + (\dot{H}_y - H_z\omega_x + H_x\omega_z)\vec{j} \\
&\quad + (\dot{H}_z - H_x\omega_y + H_y\omega_z)\vec{k}.
\end{aligned} \tag{2.75}$$

If it is now assumed that $(\vec{i}, \vec{j}, \vec{k})$ are the principal axes of the body (i.e. all of the products of inertia about these axes are zero) we can write,

$$\vec{H}_O = (I_{xx_O}\omega_x)\vec{i} + (I_{yy_O}\omega_y)\vec{j} + (I_{zz_O}\omega_z)\vec{k} \tag{2.76}$$

---

8 Leonhard Euler (1707–1783). His book *Mechanica* (1736), presented Newtonian dynamics in the form of mathematical analysis for the first time. He went on to extend Newton's laws of motion to include the dynamics of rigid bodies. Euler introduced the notation $f(x)$ for a function, e for the base of natural logarithms, i for the $\sqrt{-1}$, $\sum$ for summation, the notation for finite differences Δy and $\Delta^2 y$ and many others.

and then,

$$\dot{H}_x = I_{xx_O}\dot{\omega}_x \tag{2.77}$$

$$\dot{H}_y = I_{yy_O}\dot{\omega}_y$$

$$\dot{H}_z = I_{zz_O}\dot{\omega}_z \tag{2.78}$$

so that, from Equation 2.75,

$$\frac{d\vec{H}_O}{dt} = (I_{xx_O}\dot{\omega}_x - I_{yy_O}\omega_y\omega_z + I_{zz_O}\omega_y\omega_z)$$

$$+ (I_{yy_O}\dot{\omega}_y - I_{zz_O}\omega_x\omega_z + I_{xx_O}\omega_x\omega_z)$$

$$+ (I_{zz_O}\dot{\omega}_z - I_{xx_O}\omega_x\omega_y + I_{yy_O}\omega_x\omega_y). \tag{2.79}$$

Finally, if we equate the total externally applied moment about O ($\vec{M}_O = M_{x_O}\vec{i} + M_{y_O}\vec{j} + M_{z_O}\vec{k}$) to $\frac{d\vec{H}_O}{dt}$ from Equation 2.79 we get,

$$M_{x_O} = I_{xx_O}\dot{\omega}_x - (I_{yy_O} - I_{zz_O})\omega_y\omega_z$$

$$M_{y_O} = I_{yy_O}\dot{\omega}_y - (I_{zz_O} - I_{xx_O})\omega_x\omega_z$$

$$M_{z_O} = I_{zz_O}\dot{\omega}_z - (I_{xx_O} - I_{yy_O})\omega_x\omega_y. \tag{2.80}$$

The moment balance equations (Equation 2.80) are known as Euler's equations. The assumptions built into them are,

1. They are only applicable if x–y–z are body-fixed principal axes of inertia.
2. The reference point O must be either the center of mass G or a fixed point.

2.12 Throwing a Spiral

If we consider axisymmetric bodies, the products of inertia are zero by virtue of the symmetry and Euler's equations are readily applicable. In the case of bodies with no external moments applied, Euler's equations become very useful and interesting. Most modern textbooks take up the subject of symmetric satellites at this point but, to keep this description more down to earth, we look at the problem of throwing a perfect spiral with an American style football.

Figure 2.9 shows the situation to be considered. The quarterback has thrown the football and it is in flight. We describe its motion using the body-fixed reference frame $(\vec{i}, \vec{j}, \vec{k})$ shown in the figure. We take the reference point to be the center of mass of the football, which is, by symmetry, located at its geometric center. The moments of inertia are unknown but, again by symmetry, we know that $I_{yy_G} = I_{zz_G}$ and that I_{xx_G} is different. Let $I_{xx_G} = I_1$ and $I_{yy_G} = I_{zz_G} = I_2$.

Since there is no externally applied moment, $M_{G_x} = M_{G_y} = M_{G_z} = 0$, and we can write Euler's equations (Equation 2.80) for this case as,

$$0 = I_1\dot{\omega}_x - (I_2 - I_2)\omega_y\omega_z$$

$$0 = I_2\dot{\omega}_y - (I_2 - I_1)\omega_x\omega_z$$

$$0 = I_2\dot{\omega}_z - (I_1 - I_2)\omega_x\omega_y. \tag{2.81}$$

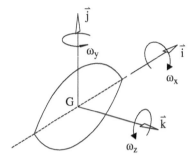

Figure 2.9 An American football in flight.

The first relationship in Equation 2.81 can be simplified to,

$$\dot{\omega}_x = 0.$$

This simply says that the angular velocity about the *x*-axis will not change with time. In other words, when the football leaves the hands of the quarterback, the rate of spin about the long axis through the ball is fixed. Since the quarterback intends to throw a "spiral", this angular velocity is usually relatively large. We can define the constant angular velocity about the *x*-axis to be $\omega_x = \Omega$.

The second and third relationships in Equation 2.81 speak to the development of angular velocities about the short axes of the football. A true spiral will have an angular velocity only about the long axis (i.e. $\Omega \neq 0, \omega_y = \omega_z = 0$).

The second and third relationships can be simplified to,

$$\dot{\omega}_y = \left(\frac{I_2 - I_1}{I_2}\right)\Omega\,\omega_z$$

$$\dot{\omega}_z = \left(\frac{I_1 - I_2}{I_2}\right)\Omega\,\omega_y.$$

If the initial values of the angular velocities about the short axes are zero (i.e. when the ball leaves the hand of the quarterback $\omega_y = \omega_z = 0$), then these two equations show that the rates of change of ω_y and ω_y will be zero and angular velocities about the short axes will never develop. In other words, the football will spin only about its long axis and the pass is a perfect spiral. If, on the other hand, the quarterback introduces a slight "wobble" by having an initial non-zero value of either ω_y or ω_z, then the rates of change of both ω_y and ω_z will be non-zero as time progresses and the pass will not be a spiral.

2.13 A Two Body System

Figure 2.10 shows a wedge of mass *M* resting on a frictionless horizontal surface. A cylinder (radius = *R*, mass = *m*) is resting on the inclined surface of the wedge. A spring of stiffness *k* and rest length ℓ_0 connects the center of the cylinder to a point on the wedge. The spring is parallel to the inclined surface. There is Coulomb friction between the cylinder and the wedge. The coefficient of friction is μ.

The equations of motion for the system are desired.

Figure 2.10 A cylinder on a wedge.

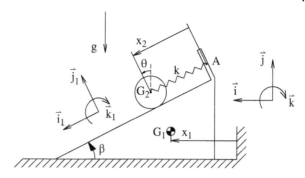

Section 2.4 gave a step by step procedure for deriving equations of motion. We will employ that procedure here.

1. *Kinematics – choose the coordinates (also known as degrees of freedom) to be used to describe the motion and derive expressions for the absolute velocities and accelerations of the masses under consideration.*

   We begin by choosing the degrees of freedom of the system. Degrees of freedom are coordinates that, when given values, will specify the translational and rotational position of each body in the system. It is often helpful to consider making a drawing of the system as it moves and asking yourself how many coordinates (degrees of freedom) you would need to make the drawing. In this case, there are three degrees of freedom.

   The wedge can only translate horizontally on the frictionless surface – use x_1 to describe the horizontal motion of the wedge away from its initial position. Clearly, if we know where the wedge was initially and we know how far it has translated horizontally from that position, we can draw the wedge. x_1 is the first degree of freedom.

   The center of the cylinder can move up and down the inclined surface of the wedge. As it does, the length of the spring changes. If we use the variable x_2 to denote the length of the spring, we can determine both the force in the spring and the position of the center of the cylinder relative to a point on the wedge. x_2 is the second degree of freedom.

   Since the cylinder is perfectly round and appears to have no distinguishing marks on its surface, the third degree of freedom is not quite as obvious. Imagine though that the end of the cylinder had a mark that we could follow as the cylinder rotated. At some times the mark would be at the top of the cylinder. At other times it would be at the bottom. We could not draw the system unless we knew how far the cylinder had rotated from its initial position. The third degree of freedom is the rotation of the cylinder. Let this be θ. x_1, x_2, θ are shown in their assumed positive senses on Figure 2.10.

   Next, we need to derive kinematic expressions for the absolute acceleration of the various bodies. In the end we will write differential equations of motion related to each degree of freedom so we will, at a minimum, need the absolute accelerations in the directions of the degrees of freedom. That is, we will need the horizontal acceleration of the wedge since it has already been decided that it can only move horizontally. We will need the translational acceleration of the center of mass of the cylinder in a direction aligned with the inclined plane on the wedge because of the definition of x_2. Finally, we will need an expression for the angular acceleration of the cylinder in order to write an equation of motion for the angular coordinate θ.

The accelerations can be expressed as follows.

The horizontal acceleration of the wedge is found by successive differentiations starting from the position vector, as follows.

$$\vec{p}_{G_1} = x_1 \vec{i} \tag{2.82}$$

$$\vec{v}_{G_1} = \dot{x}_1 \vec{i} \tag{2.83}$$

$$\vec{a}_{G_1} = \ddot{x}_1 \vec{i}. \tag{2.84}$$

The absolute acceleration of the center of mass of the cylinder can be developed from the position vector as shown in Equation 2.85. Two facts used in the development are,

- $\vec{a}_A = \vec{a}_{G_1}$ since all points on the wedge have the same motion as it translates horizontally.
- The unit vectors are related by the constant angle β as follows.

$$\vec{i} = \cos \beta \vec{i}_1 + \sin \beta \vec{j}_1$$

$$\vec{j} = -\sin \beta \vec{i}_1 + \cos \beta \vec{j}_1$$

$$\vec{p}_{G_2} = \vec{p}_A + \vec{p}_{G_2/A}$$
$$\vec{p}_{G_2} = \vec{p}_A + x_2 \vec{i}_1$$
$$\vec{v}_{G_2} = \vec{v}_A + \dot{x}_2 \vec{i}_1$$
$$\vec{a}_{G_2} = \vec{a}_A + \ddot{x}_2 \vec{i}_1$$
$$\vec{a}_{G_2} = \ddot{x}_1 \vec{i} + \ddot{x}_2 \vec{i}_1$$
$$\vec{a}_{G_2} = \ddot{x}_1 (\cos \beta \, \vec{i}_1 + \sin \beta \, \vec{j}_1) + \ddot{x}_2 \vec{i}_1$$
$$\vec{a}_{G_2} = (\ddot{x}_1 \cos \beta + \ddot{x}_2) \vec{i}_1 + \ddot{x}_1 \sin \beta \, \vec{j}_1. \tag{2.85}$$

The angular acceleration of the cylinder is found from,

$$\vec{\omega}_{cyl} = -\dot{\theta} \vec{k}$$
$$\vec{\alpha}_{cyl} = -\ddot{\theta} \vec{k}. \tag{2.86}$$

Note that the angle θ is defined to be positive in the negative $\vec{k}$ direction so that its angular velocity and angular acceleration are also in the negative $\vec{k}$ direction.

2. *Free body diagrams (FBDs) – sketch the masses under consideration as if they are in space with no forces acting on them. Then add to the sketches all of the externally applied forces acting on the masses. Also add the internal forces of interaction between the masses, being sure that they act in equal and opposite pairs as stipulated by Newton's third law. The FBDs should show the positive sense of the accelerations derived in step 1.*

Figure 2.11 shows the free body diagrams of the cylinder and the wedge. The forces shown are,

Figure 2.11 FBDs of the cylinder and the wedge.

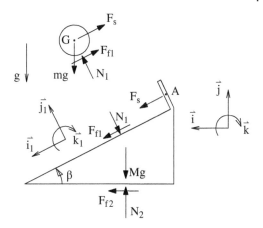

- N_1 – the normal force acting between the cylinder and the wedge. By definition, this force must be perpendicular to the plane of the wedge. Its magnitude is unknown.
- N_2 – the normal force between the wedge and the supporting surface. Again, the direction is known but the magnitude is not. This force is not, in fact, required for the analysis being done here since we are only interested in horizontal motions of the wedge, but is included for completeness.
- F_{f_1} – the frictional force acting between the cylinder and the wedge. The direction is known to be perpendicular to the normal force and in opposition to the relative velocity between the two bodies at the point of contact if slip is occurring. If there is slip, the magnitude of the friction force is $\pm \mu N_1$ – if there is no slip, the magnitude of the friction force lies between $-\mu N_1$ and $+\mu N_1$ and acts in the direction required to enforce the "no-slip" condition.
- F_{f_2} – the frictional force acting between the wedge and the supporting surface is included for completeness. It is zero in this problem because the contact was designated as frictionless.
- F_s – this is the force in the spring connecting the cylinder to the wedge. The spring stiffness was given as k, and the undeflected length as ℓ_0. In order to be sure that the sign of the force is correct as we develop an expression for it, we must refer to the assumed direction shown on the FBD. Figure 2.11 shows that the spring was assumed to be in tension (i.e. the forces are acting to pull the two bodies together). If this is the case, then it must also be assumed that the spring is longer than its undeflected length (i.e. $x_2 > \ell_0$). The displacement of the spring is therefore $\delta = x_2 - \ell_0$ and the force in the spring, acting in the direction on the FBD, is $F_s = k\delta = k(x_2 - \ell_0)$.
- mg and Mg – these are the weights of the two bodies acting vertically downward.

3. *Force Balance Equations (Newton's second law) – using the FBDs, equate the vector sum of forces on each body to its mass multiplied by its vector acceleration and the vector sum of moments on each body to its moment of inertia multiplied by its angular acceleration[9]. Referring to Figure 2.11, we write the force and moment balance equations as follows.*

---

9 In general we should use $\vec{M}_G = \frac{d\vec{H}_G}{dt}$ but, in two dimensional cases, this is often simplified to $\vec{M}_G = I_G\vec{\alpha}$.

First we recognize that we simply require a horizontal force balance on the wedge, leading to,

$$\sum F_{\vec{i}} \; : \; M\ddot{x}_1 = F_{f_1} \cos \beta - N_1 \sin \beta + F_s \cos \beta. \tag{2.87}$$

Next, the cylinder is accelerating in both the $\vec{i}_1$ and $\vec{j}_1$ (see Equation 2.85) directions so that two force balances are required. They are,

$$\sum F_{\vec{i}_1} \; : \; m(\ddot{x}_1 \cos \beta + \ddot{x}_2) = -F_{f_1} - F_s + mg \sin \beta \tag{2.88}$$

$$\sum F_{\vec{j}_1} \; : \; m\ddot{x}_1 \sin \beta = N_1 - mg \cos \beta. \tag{2.89}$$

Finally, the angular acceleration of the cylinder requires that we write,

$$\sum M_{\vec{k}_1} \; : \; -I\ddot{\theta} = -F_{f_1} R. \tag{2.90}$$

4. *Manipulate and solve the equations – the first three steps will lead to a set of equations with a set of unknowns. All that is left is to manipulate and combine the equations in order to extract the desired solution.*

   We first count the number of equations and the number of unknowns to see if we need further information before attempting a solution. The full set of equations available to us are five in total.

$$M\ddot{x}_1 = F_{f_1} \cos \beta - N_1 \sin \beta + F_s \cos \beta$$
$$m(\ddot{x}_1 \cos \beta + \ddot{x}_2) = -F_{f_1} - F_s + mg \sin \beta$$
$$m\ddot{x}_1 \sin \beta = N_1 - mg \cos \beta$$
$$-I\ddot{\theta} = -F_{f_1} R$$
$$F_s = k(x_2 - \ell_0).$$

Counting the unknowns, we find that there are the following six.
- x_1 and its derivatives[10]
- x_2 and its derivatives
- θ and its derivatives
- F_{f_1}
- N_1
- F_s.

Given five equations with six unknowns means we need to find another relationship between variables before we have a chance of generating a solution. In fact, we have already discussed the relationship required – *we know how the frictional force between the cylinder and the wedge behaves.*

We have two alternatives. If the cylinder rolls without slipping, F_{f_1} is whatever it must be to enforce the "no-slip" condition and we can generate a kinematic relationship between the velocity of the center of the cylinder and its angular velocity.

---

10 A variable and its derivatives are counted as a single unknown since they are related by differential equations we do not derive. For example, $\dot{x} = \frac{dx}{dt}$ goes without saying.

If the cylinder rolls and slips, then there is relative motion at the point of contact between the two bodies and the frictional force is known to be at its limiting value, $F_{f_1} = \mu N_1$, with a direction opposing the relative motion.

Consider first the case where the cylinder rolls without slipping. Define the point of contact between the cylinder and the wedge to be point B. The definition of the "no-slip" condition is that point B must have the same velocity whether we consider it a point on the wedge or a point on the cylinder.

If point B is on the wedge, we know its velocity is the same as that of any other point on the wedge because the wedge simply translates. That is,

$$\vec{v}_B^{\text{wedge}} = \dot{x}_1 \vec{i}. \tag{2.91}$$

If point B is on the cylinder, we find its velocity as follows.

$$\vec{p}_B^{\text{cyl}} = \vec{p}_A + \vec{p}_{B/A}$$

$$\vec{p}_B^{\text{cyl}} = \vec{p}_A + x_2 \vec{i}_1 - R\vec{j}_2$$

$$\vec{v}_B^{\text{cyl}} = \vec{v}_A + \dot{x}_2 \vec{i}_1 - (-\dot{\theta}\vec{k}_2) \times (R\vec{j}_2)$$

$$\vec{v}_B^{\text{cyl}} = \vec{v}_A + \dot{x}_2 \vec{i}_1 - R\dot{\theta}\vec{i}_2$$

$$\vec{v}_B^{\text{cyl}} = \dot{x}_1 \vec{i} + \dot{x}_2 \vec{i}_1 - R\dot{\theta}\vec{i}_2$$

$$\vec{v}_B^{\text{cyl}} = \dot{x}_1 \vec{i} + (\dot{x}_2 - R\dot{\theta})\vec{i}_1. \tag{2.92}$$

What was just done needs explanation. Defining the position of point B in the cylinder requires that the relative position of two points in a rigid body (the cylinder) be defined. The relative velocity of these two points (G_2 and B) can only be due to the angular velocity of the coordinate system in which they are expressed. Clearly, the two coordinate systems used in the analysis, $(\vec{i}, \vec{j}, \vec{k})$ and $(\vec{i}_1, \vec{j}_1, \vec{k}_1)$, do not rotate and are incapable of expressing $\vec{v}_{B/G_2}$. It was therefore necessary to introduce the coordinate system $(\vec{i}_2, \vec{j}_2, \vec{k}_2)$, which is fixed in the cylinder and has angular velocity $-\dot{\theta}\vec{k}_2$. $(\vec{i}_2, \vec{j}_2, \vec{k}_2)$ is aligned with $(\vec{i}_1, \vec{j}_1, \vec{k}_1)$ at the instant being considered so the final result has been written in the $(\vec{i}_1, \vec{j}_1, \vec{k}_1)$ system. Equating the two expressions for $\vec{v}_B$ from Equations 2.91 and 2.92 gives,

$$\vec{v}_B^{\text{wedge}} = \vec{v}_B^{\text{cyl}}$$

$$\dot{x}_1 \vec{i} = \dot{x}_1 \vec{i} + (\dot{x}_2 - R\dot{\theta})\vec{i}_1$$

$$0 = (\dot{x}_2 - R\dot{\theta})\vec{i}_1$$

$$\dot{x}_2 = R\dot{\theta}. \tag{2.93}$$

The result from Equation 2.93 becomes the sixth equation required for a solution. We then combine the equations and eliminate the variables F_{f_1}, N_1, F_s, and θ and its derivatives to get,

$$\dot{\theta} = \frac{1}{R}(\dot{x}_2)$$

$$F_{f_1} = \frac{I}{R}(\ddot{\theta}) = \frac{I}{R}\left(\frac{\ddot{x}_2}{R}\right) = \frac{I}{R^2}(\ddot{x}_2)$$

$$N_1 = m\ddot{x}_1 \sin\beta + mg\cos\beta$$

$$F_s = k(x_2 - \ell_0) \tag{2.94}$$

and, finally,

$$
\begin{bmatrix}
(M + m \sin^2\beta) & -\frac{I}{R^2}\cos\beta \\
m\cos\beta & \left(m + \frac{I}{R^2}\right)
\end{bmatrix}
\begin{Bmatrix}
\ddot{x}_1 \\
\ddot{x}_2
\end{Bmatrix}
=
$$

$$
\begin{Bmatrix}
k(x_2 - \ell_0)\cos\beta - mg\cos\beta\sin\beta \\
mg\sin\beta - k(x_2 - \ell_0)
\end{Bmatrix}.
\tag{2.95}
$$

The end result is a set of two, simultaneous, nonlinear, second order, differential equations in the two variables x_1 and x_2. Later, we will discuss methods for solving the differential equations arising from derivations such as this. For the time being, we must simply realize that there is a significant assumption built into these equations. Namely that the cylinder rolls without slipping on the wedge. Any solution to the equations must continually check to see if the required value of F_{f_1} exceeds μN_1. If it does, these differential equations are no longer applicable and we must derive the equations of motion corresponding to $F_{f_1} = \mu N_1$ and use them for the solution.

2.14 Gyroscopic Motion

Gyroscopic motion occurs whenever the axis about which a body is spinning is itself rotating about another axis. Gyroscopic effects hold a certain mystique because they never behave in the way everyday experience would tell people they should.

In fact, gyroscopic effects can be readily understood from the equations we have already developed and it is the purpose of this section to give the reader that understanding without unduly complicating matters with equations. We seek a feeling for gyroscopic effects that comes from an understanding of the rate of change of direction of a vector – in this case, the angular momentum vector.

A moment balance is always involved in gyroscopic considerations. The terms involving the rate of change of direction of the angular momentum vector are the "gyroscopic terms" in the equations of motion.

$$
\sum \vec{M}_O = \frac{d\vec{H}_O}{dt} = \dot{\vec{H}}_O + \underbrace{\vec{\omega} \times \vec{H}_O}_{\substack{\text{gyroscopic} \\ \text{terms}}}.
\tag{2.96}
$$

Equation 2.59 gives a general expression for the angular momentum vector, either about a fixed point or about the center of mass of the body. If we restrict the discussion to axi-symmetric bodies and use the center of mass G as the reference point, the products of inertia will vanish and the angular momentum vector will simply become,

$$
\vec{H}_G = I_{xx_G}\omega_x\vec{i} + I_{yy_G}\omega_y\vec{j} + I_{zz_G}\omega_z\vec{k}
\tag{2.97}
$$

where the axes $(\vec{i}, \vec{j}, \vec{k})$ are fixed in the rigid body.

Further, if we assume that one component of the angular velocity vector is significantly larger than the other two (say, for example, $\omega_x \gg \omega_y$ and $\omega_x \gg \omega_z$) then we can approximate the angular momentum vector as,

$$\vec{H}_G \approx I_{xx_G}\omega_x\vec{i} \tag{2.98}$$

or,

$$\vec{H} \approx I\vec{\omega} \tag{2.99}$$

where the subscripts have been dropped to show that the angular momentum of the body is essentially aligned with its "spin axis". That is, the axis about which the angular speed is the greatest.

It is easy to think of many examples where there is a dominant component to the angular velocity vector. A top spinning on a horizontal surface, for instance, has an angular velocity vector and a resulting angular momentum vector that are directed primarily in the vertical direction. The top may well be wobbling but the components of the angular velocity about directions other than the vertical will be relatively small.

Another example is a bicycle. If we consider the bicycle to be made of four components – the front forks, the frame, and two wheels, only the wheels rotate as the bicycle moves. A bicycle moving forward will have a large angular momentum vector generated by each wheel and the angular momentum vectors can be visualized as pointing out of the wheel centers to the left[11]. The angular momentum vectors will get longer as the bicycle goes faster but the direction will not change.

There are numerous other examples of systems with rapidly rotating components that generate significant angular momentum vectors. Think of the engines in cars that rotate rapidly about fixed directions in the car. Think of jet aircraft whose engines have very high rates of rotation and always point the same way relative to the plane.

Figure 2.12 shows a top spinning on a horizontal surface. The coordinate system is fixed in the body of the top and therefore spins with it, having an angular velocity $\vec{\omega} = \omega\vec{k}$. The

Figure 2.12 A spinning top.

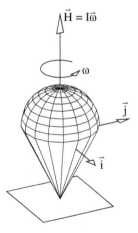

---

11 Check this statement by visualizing the direction in which the wheels turn and using the right hand rule to determine the direction of the angular momentum vector.

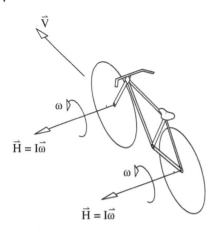

Figure 2.13 A bicycle.

moment of inertia of the top about the spin direction is *I* so that the angular momentum vector is $\vec{H} = I\vec{\omega}$. From the figure it is clear that the angular momentum vector is aligned with the angular velocity vector and is oriented vertically upward as shown[12].

Figure 2.13 shows the situation for a bicycle that is being ridden in a straight line. The angular momentum vectors of the two wheels are shown pointing horizontally to the left of the bicycle. In fact, the moment of inertia of a bicycle wheel is relatively large and the angular velocity grows with forward speed so that the magnitude of the angular velocity vectors is large. The reader can imagine that the equivalent vectors for a motorcycle traveling at high speeds will be very large indeed. The total angular momentum vector of the bicycle is the sum of the two vectors shown in Figure 2.13.

We now turn our attention to the action that we would be required to take in order to reorient the top or the bicycle.

Consider first the top. Figure 2.14 shows the top after someone has tipped it slightly so that the point *A* has moved vertically downward. We can assume that the change in orientation was caused by a force applied vertically downward somewhere on the periphery of the top. The question is *"where do you apply the force?"*.

Quite clearly, if the top were not spinning a force applied vertically downward at point *A* would do the trick. However, since there is a significant angular momentum vector, we have to take account of the gyroscopic effect[13]. Figure 2.14 shows the angular momentum vector both before ($\vec{H}_0$) and after ($\vec{H}_1$) the tipping takes place. The magnitude of the vector is not likely to change significantly as it tips. The major change in the angular momentum vector is a directional change. The directional change is indicated by the vector $\Delta\vec{H}$ shown in the figure.

$\Delta\vec{H}$ is aligned with the unit vector $\vec{i}$ in Figure 2.14 since the point *A* lies on that axis. If the change in the orientation of the top occurred over a period of time Δt, then $\dfrac{d\vec{H}}{dt}$ can

---

12 The angular velocity vector of the top is that which would be imparted if it were thrown by a left-handed person. Right-handed readers would spin it in the opposite direction but that would mean that the angular momentum vector would point downward and the figure wouldn't be nearly so clear.

13 It is the angular momentum vector and its resistance to changing direction that causes the top to stand upright in the first place.

Figure 2.14 The "tipped" top.

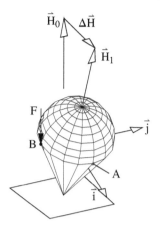

be approximated by $\dfrac{\Delta \vec{H}}{\Delta t}$ and, since Δt is a scalar, the direction of $\dfrac{\Delta \vec{H}}{\Delta t}$ is the same as the direction of $\Delta \vec{H}$ and the moment applied to the top must also be in this direction since $\vec{M} = \dfrac{d\vec{H}}{dt}$.

By the right hand rule, a positive moment in the $\vec{i}$ direction would require a vertical force, F, to be applied, not at point A since there would be no moment about the $\vec{i}$ axis in that case, but at a point like B that has a moment arm that will give rise to a moment about the $\vec{i}$ axis. In fact, the maximum moment arm on the axisymmetric top occurs when the force is applied at a point B located on the negative $\vec{j}$ axis. Tops always tip 90° away from the point of application of a vertical force on their periphery. This is easily demonstrated experimentally with a simple top on a tabletop.

The fact that the response to an action occurs 90° away from the action is a basic property of gyroscopic effects. An analyst, using basic geometrical information about the angular velocity vector and its directional rate of change can easily determine what to expect from a system with significant rotating masses.

Next, we consider the bicycle and what it would take to reorient it so that it will move in a different direction on the horizontal plane. In other words, we would like to steer the bicycle so that it turns away from the straight path it is following in Figure 2.13. Let us consider the case where we would like the bicycle to turn to the right. Figure 2.15 shows the effect we must have on the angular momentum vectors in order to do this. The angular momentum vectors are still perpendicular to the plane of the rotating wheels but they have turned with the bicycle and, as a result, there must have been a directional rate of change of the vectors. The original angular momentum vectors $(\vec{H}_0)$ and the rotated vectors $(\vec{H}_1)$ as well as the change in the vectors $(\Delta \vec{H} = \vec{H}_1 - \vec{H}_0)$ are shown on Figure 2.15. As we know, the rate of change of the angular momentum vector (approximately equal to the total change $\Delta \vec{H}$ divided by the scalar length of time Δt over which the change occurred) must be equal to the applied moment both in magnitude and direction. Leaving aside the magnitude, the direction of the applied moment is easily seen to be aligned with the forward motion of the bicycle. By the right hand rule, this means that a vertical force must have been applied to the right or left of the centerline of the bicycle. The most easily applied force is that from

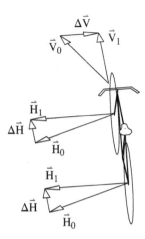

Figure 2.15 The bicycle turning to the right.

shifting the weight of the rider to the right so that the vertically downward gravity force on the rider causes the required moment.

Any reader who has ever ridden a bicycle "no-hands" will know that shifting the rider's weight to the right or left is sufficient to steer a bicycle accurately along the chosen path.

Exercises

Descriptions of the systems referred to in the Exercises are contained in Appendix A.

2.1 Draw FBDs, write force and moment balance equations, and derive the equations of motion for system 1. Show that the force acting on body OA due to its interaction with the pin at A is

$$\vec{R}_A = m_2[-g\sin\theta_1 + d_1\dot{\theta}_1^2 + d_2(\ddot{\theta}_1 + \ddot{\theta}_2)\sin\theta_2 + d_2(\dot{\theta}_1 + \dot{\theta}_2)^2\cos\theta_2]\vec{i}$$
$$+ m_2[-g\cos\theta_1 - d_1\ddot{\theta}_1 - d_2(\ddot{\theta}_1 + \ddot{\theta}_2)\cos\theta_2 + d_2(\dot{\theta}_1 + \dot{\theta}_2)^2\sin\theta_2]\vec{j}.$$

2.2 Draw FBDs, write force and moment balance equations, and derive the equations of motion for system 2. Show that the tension in the string BC can be expressed as

$$T = mg\cos(\theta - \phi) - m\ddot{x}\sin\phi + m\dot{\theta}^2(x - \ell)\sin\phi$$
$$- 2m\dot{x}\dot{\theta}\cos\phi + md(\dot{\theta} - \dot{\phi})^2 + m(\ell - x)\ddot{\theta}\cos\phi.$$

2.3 Show that the normal force that the slot exerts on the mass in system 4 has magnitude

$$N = m(\dot{\Omega}x + 2\Omega\dot{x})$$

and indicate in which direction it acts.

2.4 Derive the equations of motion for system 5. Show that the friction force acting on the oil drum can be expressed as

$$F_f = mg \sin \theta - m(d - x)\dot{\theta}^2 - mr\ddot{\theta} - m\ddot{x}$$

and indicate in which direction it acts.

2.5 Consider the particle of mass m and the rigid rod of system 6 to be a single rigid body. Where is the center of mass of the body? Draw an FBD and explain why there are five unknown reaction components at point A. What is the angular momentum of the body about the center of mass?

2.6 Show that the equation of motion governing the angle θ in system 6 is

$$mr\ddot{\theta} - m\omega^2(d - r\cos\theta)\sin\theta + mg\cos\theta = 0.$$

2.7 Draw FBDs, write force and moment balance equations, and derive the equation of motion for system 7. Show that the component of the reaction force acting on rod OA at point O in the direction from O to B can be expressed as

$$F_O = F - 8mL\ddot{\theta}\sin\theta - 8mL\dot{\theta}^2\cos\theta.$$

2.8 Derive the equations of motion for system 16 and show that the reaction force acting on the rod is

$$R_A = mg\sin\theta + mr\dot{\theta}^2 + m\ell(\ddot{\theta} + \ddot{\phi})\sin\phi + m\ell(\dot{\theta} + \dot{\phi})^2\cos\phi.$$

2.9 Show that the angular momentum of the rectangular mass of system 22 about its center of mass and expressed in a right-handed set of body fixed principal axes where $\vec{\imath}$ is aligned with AB and $\vec{k}$ is aligned with BC is

$$\vec{H}_G = \frac{1}{6}mb^2\left(\omega_1\vec{\imath} + \frac{5}{2}\omega_0\sin\theta\vec{\jmath} + \frac{5}{2}\omega_0\cos\theta\vec{k}\right).$$

2.10 Consider system 22 in the case where ω_0 and ω_1 are both constant. Show that the moment that must be applied about the center of mass of the rectangular body is

$$\vec{M}_G = \frac{1}{4}mb^2\omega_0\omega_1(-\cos\theta\vec{\jmath} + \sin\theta\vec{k})$$

where the unit vectors are the same set used in Exercise 2.9.

2.11 Use Newton's Laws to derive the equations of motion for system 23.

2.12 Using the angle, θ, as the single degree of freedom, show that the equation-of-motion of the rod in system 12 can be written as

$$\frac{4}{3}md^2\ddot{\theta} + 4kd^2\sin\theta\cos\theta - mgd\cos\theta = 0.$$

3

Lagrange's Equations of Motion

In this chapter, we consider the development of Lagrange's[1] equations of motion. We start with Newton's laws and work from them to derive a method of writing equations of motion by taking derivatives of scalar expressions for kinetic and potential energy in the system.

Use of Lagrange's equations gives the analyst two distinct advantages when deriving the equations of motion. First, the vector kinematic analysis is shorter than it is with a direct application of Newton's laws since acceleration vectors need not be found. This is because the kinetic and potential energy expressions can be derived from velocity vectors and position vectors respectively. Secondly, there is no need to draw free body diagrams for each of the rigid bodies in the system because the forces of constraint between the bodies do no work and are therefore not required for the analysis.

Of course, there are also disadvantages. The method requires a great deal of differentiation, sometimes of relatively complicated functions. Some analysts prefer the kinematics of Newton's method over the differentiation involved in Lagrange's equations. Some point to a lack of physical feeling for problems without free body diagrams as being a disadvantage of the method. Finally, if the intent of analyzing the dynamics of a system is to predict loads, which could be carried forward into a structural analysis for instance, the forces of interaction between bodies are not available from a straightforward application of Lagrange's equations.

3.1 An Example to Start

Rather than going directly into the derivation of Lagrange's equation we will start with a relatively simple example so that the form of the equation and the way it is applied can be seen.

Lagrange's equation is

$$\frac{\mathrm{d}}{\mathrm{d}t}\left(\frac{\partial T}{\partial \dot{q}}\right) - \frac{\partial T}{\partial q} + \frac{\partial U}{\partial q} = Q_q \tag{3.1}$$

---

1 Joseph-Louis Lagrange (1736–1813), an Italian/French mathematician, is well known for his work on calculus of variations, dynamics, and fluid mechanics. In 1788 Lagrange published the *Mécanique Analytique* summarizing all the work done in the field of mechanics since the time of Newton, thereby transforming mechanics into a branch of mathematical analysis.

The Practice of Engineering Dynamics, First Edition. Ronald J. Anderson.
© 2020 John Wiley & Sons Ltd. Published 2020 by John Wiley & Sons Ltd.
Companion Website: www.wiley.com/go/anderson/engineeringdynamics

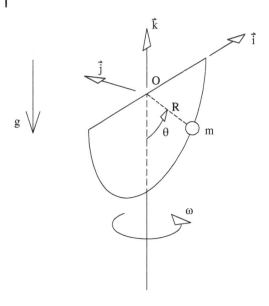

Figure 3.1 A mass on a wire.

where

- T = the total kinetic energy of the system
- U = the total potential energy of the system
- q = a generalized coordinate
- $\dot{q}$ = the time derivative of q
- Q_q = the generalized force corresponding to a variation of q.

The reader will certainly have some feeling for the kinetic and potential energy of a system but probably not for the concepts of generalized coordinate and generalized force. The full definitions of these terms will be given later. At this point we will simply apply Lagrange's equation to a system as if all were understood.

Figure 3.1 shows a small mass, m, which slides on a semi-circular wire that rotates about a vertical axis. The wire has radius R. Gravity acts to pull the mass to the bottom of the semi-circle while centripetal effects try to move it to the top. The equation of motion governing the single degree of freedom, θ, is desired.

We first determine the kinetic energy of the system. This requires that we find an expression for the absolute velocity of the mass. Using the rotating coordinate system shown in the figure, we can write

$$\vec{p}_{m/O} = R\sin\theta\,\vec{i} - R\cos\theta\,\vec{k}. \tag{3.2}$$

The absolute velocity is

$$\vec{v}_m = \vec{v}_O + \frac{d}{dt}\vec{p}_{m/O}. \tag{3.3}$$

Then, recognizing that $\vec{v}_O = 0$ and that $\dot{R} = 0$,

$$\vec{v}_m = (R\dot{\theta}\cos\theta\,\vec{i} + R\dot{\theta}\sin\theta\,\vec{k}) + \omega\vec{k} \times (R\sin\theta\,\vec{i} - R\cos\theta\,\vec{k}) \tag{3.4}$$

which can be simplified to,

$$\vec{v}_m = R\dot{\theta}\cos\theta\vec{i} + \omega R\sin\theta\vec{j} + R\dot{\theta}\sin\theta\vec{k}. \tag{3.5}$$

The kinetic energy of the system is then

$$T = \frac{1}{2}m(\vec{v}_m \cdot \vec{v}_m) \tag{3.6}$$

which becomes, after substitution of Equation 3.5 and some simplification,

$$T = \frac{1}{2}mR^2(\dot{\theta}^2 + \omega^2\sin^2\theta). \tag{3.7}$$

The potential energy of the system is due to gravity only. If the datum for potential energy is taken to be at point O, the potential energy, U, of the system is simply

$$U = -mgR\cos\theta. \tag{3.8}$$

Having expressions for T and U and a single degree of freedom, θ, we can apply Lagrange's equation (Equation 3.1) and find

$$q = \theta$$
$$\dot{q} = \dot{\theta}$$
$$\frac{\partial T}{\partial\dot{\theta}} = mR^2\dot{\theta}$$
$$\frac{d}{dt}\left(\frac{\partial T}{\partial\dot{\theta}}\right) = mR^2\ddot{\theta}$$
$$\frac{\partial T}{\partial\theta} = mR^2\omega^2\sin\theta\cos\theta$$
$$\frac{\partial U}{\partial\theta} = mgR\sin\theta$$
$$Q_\theta = 0. \tag{3.9}$$

Substituting the expressions from Equation 3.9 into Lagrange's equation gives the desired equation of motion

$$mR^2\ddot{\theta} - mR^2\omega^2\sin\theta\cos\theta + mgR\sin\theta = 0. \tag{3.10}$$

Clearly, Equation 3.10 could be further simplified by factoring out the group mR but this would take away the ability to look at the individual terms and give a physical explanation for them. Whenever an equation is derived, the first test for correctness is to see if all of the terms have the same dimensions. In this case, the first term has dimensions of ML^2/T^2 where M is *mass*, L is *length*, and T is *time*. Note that angles such as θ are dimensionless since they are defined by an arc length divided by a radius. It follows that trigonometric functions such as $\sin\theta$ and $\cos\theta$ are also dimensionless. Angular velocities therefore have dimensions derived from angles divided by time, $1/T$, and angular accelerations are expressed as $1/T^2$. Using these conventions, it is easy to see that all three terms in Equation 3.10 have the same dimensions[2].

---

2 Note the difference between dimensions and units. *Dimensions* refer to physical characteristics such as mass, length, or time. *Units* refer to the system of measurement we use to substitute numbers into an equation. Examples are kilograms for mass, feet for length, and minutes for time.

The dimensions of force are ML/T^2 or mass times acceleration. Taking this into account, we can see that the three terms in Equation 3.10 all have dimensions of FL or force times length. The terms are, in fact, all moments. The third term is the most obvious because it contains the gravity force mg multiplied by a moment arm of $R\sin\theta$. The moment arm is simply the horizontal distance between the mass and point O. Lagrange's equation has produced an equation of motion based on a dynamic moment balance about the stationary point O and it did so without requiring the derivation of acceleration expressions, the drawing of free body diagrams, or the production of force and moment balance relationships. This is the power of using Lagrange's equation for deriving equations of motion.

3.2 Lagrange's Equation for a Single Particle

In this section we start with Newton's laws and derive from them Lagrange's equation for a particle.

Assume there is a single particle of mass m moving in three dimensional space and its position is expressed using three Cartesian coordinates[3] x, y, and z. Associated with each Cartesian direction is a unit vector so that the position of the particle with respect to a fixed point in space can be written as $\vec{p} = x\vec{i} + y\vec{j} + z\vec{k}$. The Cartesian coordinate system is non-rotating.

Further assume that there is a total external force, $\vec{F} = F_x\vec{i} + F_y\vec{j} + F_z\vec{k}$ acting on the particle. Then we can write the equations of motion for the particle as

$$m\ddot{x} = F_x$$
$$m\ddot{y} = F_y$$
$$m\ddot{z} = F_z. \tag{3.11}$$

Now consider the work done by the externally applied forces during a completely arbitrary, infinitesimally small, displacement $\delta\vec{s}$ in three dimensional space. We call this *virtual work*.

In Cartesian coordinates, we can write $\delta\vec{s} = \delta x\vec{i} + \delta y\vec{j} + \delta z\vec{k}$. The virtual work is then found from $\delta W = \vec{F} \cdot \delta\vec{s}$ or,

$$\delta W = F_x\delta x + F_y\delta y + F_z\delta z. \tag{3.12}$$

Consider multiplying each of the expressions in Equation 3.11 by its respective component of $\delta\vec{s}$ and adding together the terms. The result is,

$$m(\ddot{x}\delta x + \ddot{y}\delta y + \ddot{z}\delta z) = F_x\delta x + F_y\delta y + F_z\delta z. \tag{3.13}$$

The right hand side of Equation 3.13 is simply the virtual work defined in Equation 3.12. The left hand side of Equation 3.13 corresponds to the change in energy in the system that arises from having work done on it. In this case, there will be a change in kinetic energy only. Any possible potential energy terms due to gravity or elasticity have been included in the total force acting on the particle.

---

3 Named for René Descartes (1596–1650), a French philosopher and mathematician whose major work, *La géométrie*, includes his application of algebra to geometry from which we now have Cartesian geometry.

There are often constraints on the motion of particles that need to be taken into account. Let us now assume that the particle in question is not free to move arbitrarily in three dimensional space but that there is some constraint acting on it so that its motion can be completely described by two "generalized coordinates", q_1 and q_2. The three Cartesian coordinates x, y, and z must be able to be derived from the two generalized coordinates. We are simply saying that, given values for q_1 and q_2, we can use algebraic equations arising from the constraints to find x, y, and z. As an example, consider the case where the particle is constrained to remain on an inclined surface. We can locate the particle by using two coordinates on the inclined surface (the generalized coordinates) and then, using the equation of the plane, we can find the three Cartesian coordinates from the two generalized coordinates. That is

$$x = x(q_1, q_2)$$
$$y = y(q_1, q_2)$$
$$z = z(q_1, q_2).$$
(3.14)

Given that the Cartesian coordinates are related to the generalized coordinates through the relationships in Equation 3.14, we can write

$$\delta x = \frac{\partial x}{\partial q_1} \delta q_1 + \frac{\partial x}{\partial q_2} \delta q_2$$
$$\delta y = \frac{\partial y}{\partial q_1} \delta q_1 + \frac{\partial y}{\partial q_2} \delta q_2$$
$$\delta z = \frac{\partial z}{\partial q_1} \delta q_1 + \frac{\partial z}{\partial q_2} \delta q_2.$$
(3.15)

The *generalized coordinates* q_1 and q_2 must be independently variable. That is, δq_1 and δq_2 must be able to be given arbitrary small values independently without violating the constraints.

We further assume that the *constraint is smooth* so that the work done by the force of constraint is zero for arbitrary δq_1 and δq_2. This is not as restrictive a condition as it may seem at first sight. Most constraints in mechanical systems act to prevent displacements of bodies with respect to each other. Consider, for example two bodies that are pinned together at a common point. The constraint is that the two bodies cannot move apart at that point and the constraint forces will ensure that this is actually the case. Since there is no displacement associated with the constraint forces, no work is done.

Under these conditions, we can substitute Equation 3.15 into Equation 3.13 to yield the following expression for the virtual work.

$$\delta W = m\left(\ddot{x}\frac{\partial x}{\partial q_1} + \ddot{y}\frac{\partial y}{\partial q_1} + \ddot{z}\frac{\partial z}{\partial q_1}\right)\delta q_1 + m\left(\ddot{x}\frac{\partial x}{\partial q_2} + \ddot{y}\frac{\partial y}{\partial q_2} + \ddot{z}\frac{\partial z}{\partial q_2}\right)\delta q_2$$
$$= \left(F_x\frac{\partial x}{\partial q_1} + F_y\frac{\partial y}{\partial q_1} + F_z\frac{\partial z}{\partial q_1}\right)\delta q_1 + \left(F_x\frac{\partial x}{\partial q_2} + F_y\frac{\partial y}{\partial q_2} + F_z\frac{\partial z}{\partial q_2}\right)\delta q_2.$$
(3.16)

Since q_1 and q_2 are independently variable, we may consider the case where $\delta q_2 = 0$ and $\delta q_1 \neq 0$. In this case, Equation 3.16 reduces to the work done during an arbitrary variation of q_1 with q_2 being held constant. In more general systems, we think of varying *one* of the generalized coordinates while holding *all others* constant.

Equation 3.16 then reduces to

$$\delta W_{q_1} = m \left(\ddot{x} \frac{\partial x}{\partial q_1} + \ddot{y} \frac{\partial y}{\partial q_1} + \ddot{z} \frac{\partial z}{\partial q_1} \right) \delta q_1$$

$$= \left(F_x \frac{\partial x}{\partial q_1} + F_y \frac{\partial y}{\partial q_1} + F_z \frac{\partial z}{\partial q_1} \right) \delta q_1. \tag{3.17}$$

We now consider the terms on the left hand side of Equation 3.17. A representative term is

$$\ddot{x} \frac{\partial x}{\partial q_1}. \tag{3.18}$$

We can derive this term from differentiating $\dot{x} \dfrac{\partial x}{\partial q_1}$ with respect to time. That is

$$\frac{\mathrm{d}}{\mathrm{d}t} \left(\dot{x} \frac{\partial x}{\partial q_1} \right) = \underbrace{\ddot{x} \frac{\partial x}{\partial q_1}}_{\text{the term}} + \dot{x} \frac{\mathrm{d}}{\mathrm{d}t} \left(\frac{\partial x}{\partial q_1} \right). \tag{3.19}$$

Equation 3.19 can be reorganized to give the expression from Equation 3.18 as two expressions, A and B, as shown below

$$\ddot{x} \frac{\partial x}{\partial q_1} = \underbrace{\frac{\mathrm{d}}{\mathrm{d}t} \left(\dot{x} \frac{\partial x}{\partial q_1} \right)}_{\text{A}} - \underbrace{\dot{x} \frac{\mathrm{d}}{\mathrm{d}t} \left(\frac{\partial x}{\partial q_1} \right)}_{\text{B}}. \tag{3.20}$$

Consider now these two expressions.

1. *Expression A*

   Since $x = x(q_1, q_2)$, it is true that

   $$\dot{x} = \frac{\partial x}{\partial q_1} \dot{q}_1 + \frac{\partial x}{\partial q_2} \dot{q}_2. \tag{3.21}$$

   Differentiating this expression with respect to $\dot{q}_1$ yields

   $$\frac{\partial \dot{x}}{\partial \dot{q}_1} = \frac{\partial x}{\partial q_1} \tag{3.22}$$

   and, therefore,

   $$\frac{\mathrm{d}}{\mathrm{d}t} \left(\dot{x} \frac{\partial x}{\partial q_1} \right) = \frac{\mathrm{d}}{\mathrm{d}t} \left(\dot{x} \frac{\partial \dot{x}}{\partial \dot{q}_1} \right). \tag{3.23}$$

2. *Expression B*

   The order of differentiation in term B can be changed to show that

   $$\frac{\mathrm{d}}{\mathrm{d}t} \left(\frac{\partial x}{\partial q_1} \right) = \frac{\partial}{\partial q_1} \left(\frac{\mathrm{d}x}{\mathrm{d}t} \right) = \frac{\partial \dot{x}}{\partial q_1}. \tag{3.24}$$

   Substituting the new expressions for A and B into Equation 3.19 gives

   $$\ddot{x} \frac{\partial x}{\partial q_1} = \frac{\mathrm{d}}{\mathrm{d}t} \left(\dot{x} \frac{\partial \dot{x}}{\partial \dot{q}_1} \right) - \dot{x} \frac{\partial \dot{x}}{\partial q_1} \tag{3.25}$$

   where the right hand side now depends on $\dot{x}$ only whereas in Equation 3.19 both x and $\dot{x}$ appeared on the right hand side.

We now define a scalar function $\psi = \dot{x}^2/2$ and take its partial derivatives with respect to $\dot{q}_1$ and q_1 to find

$$\frac{\partial \psi}{\partial \dot{q}_1} = 2\left(\frac{\dot{x}}{2}\right)\left(\frac{\partial \dot{x}}{\partial \dot{q}_1}\right) = \dot{x}\frac{\partial \dot{x}}{\partial \dot{q}_1}$$

$$\frac{\partial \psi}{\partial q_1} = 2\left(\frac{\dot{x}}{2}\right)\left(\frac{\partial \dot{x}}{\partial q_1}\right) = \dot{x}\frac{\partial \dot{x}}{\partial q_1}. \tag{3.26}$$

The terms derived in Equation 3.26 from the scalar function ψ are exactly those that appear on the right hand side of Equation 3.25 so that we can make a substitution and write

$$\dot{x}\frac{\partial x}{\partial q_1} = \frac{d}{dt}\left[\frac{\partial(\dot{x}^2/2)}{\partial \dot{q}_1}\right] - \frac{\partial(\dot{x}^2/2)}{\partial q_1}. \tag{3.27}$$

The term we were actually concerned with in Equation 3.17 was that shown in Equation 3.27 multiplied by the particle mass m. We can multiply Equation 3.27 by the scalar m and factor it inside the derivatives since it is constant. The result is

$$m\ddot{x}\frac{\partial x}{\partial q_1} = \frac{d}{dt}\left[\frac{\partial(m\dot{x}^2/2)}{\partial \dot{q}_1}\right] - \frac{\partial(m\dot{x}^2/2)}{\partial q_1}. \tag{3.28}$$

There are two similar terms in Equation 3.17. These are $m\ddot{y}\dfrac{\partial y}{\partial q_1}$ and $m\ddot{z}\dfrac{\partial z}{\partial q_1}$. We could repeat the treatment given to $m\ddot{x}\dfrac{\partial x}{\partial q_1}$ on each of these and the result would be

$$m\ddot{y}\frac{\partial y}{\partial q_1} = \frac{d}{dt}\left[\frac{\partial(m\dot{y}^2/2)}{\partial \dot{q}_1}\right] - \frac{\partial(m\dot{y}^2/2)}{\partial q_1}$$

$$m\ddot{z}\frac{\partial z}{\partial q_1} = \frac{d}{dt}\left[\frac{\partial(m\dot{z}^2/2)}{\partial \dot{q}_1}\right] - \frac{\partial(m\dot{z}^2/2)}{\partial q_1}. \tag{3.29}$$

Substituting Equations 3.28 and 3.29 into Equation 3.17 yields

$$\delta W_{q_1} = \left(\frac{d}{dt}\left\{\frac{\partial}{\partial \dot{q}_1}\left[\frac{1}{2}m(\dot{x}^2 + \dot{y}^2 + \dot{z}^2)\right]\right\} - \frac{\partial}{\partial q_1}\left\{\frac{1}{2}m(\dot{x}^2 + \dot{y}^2 + \dot{z}^2)\right\}\right)\delta q_1$$

$$= \left(F_x\frac{\partial x}{\partial q_1} + F_y\frac{\partial y}{\partial q_1} + F_z\frac{\partial z}{\partial q_1}\right)\delta q_1. \tag{3.30}$$

We can recognize the kinetic energy of the particle appearing twice in Equation 3.30. Since the absolute velocity of the particle is

$$\vec{v}_m = \dot{x}\vec{i} + \dot{y}\vec{j} + \dot{z}\vec{k} \tag{3.31}$$

then the kinetic energy of the particle is

$$T = \frac{1}{2}m(\vec{v}\cdot\vec{v}) \tag{3.32}$$

or,

$$T = \frac{1}{2}m(\dot{x}^2 + \dot{y}^2 + \dot{z}^2) \tag{3.33}$$

which is the expression seen in Equation 3.30.

We can therefore write

$$\frac{d}{dt}\left(\frac{\partial T}{\partial \dot{q}_1}\right) - \frac{\partial T}{\partial q_1} = F_x\frac{\partial x}{\partial q_1} + F_y\frac{\partial y}{\partial q_1} + F_z\frac{\partial z}{\partial q_1}. \tag{3.34}$$

The paragraph following Equation 3.16, said *"since q_1 and q_2 are independently variable, we may consider the case where $\delta q_2 = 0$ and $\delta q_1 \neq 0$. In this case, Equation (3.16) reduces to the work done during an arbitrary variation of q_1 with q_2 being held constant. In more general systems, we think of varying one of the generalized coordinates while holding all others constant."*

Therefore, we consider now the case where $\delta q_1 = 0$ and $\delta q_2 \neq 0$. Clearly, this case would follow all of the same steps as the previous one and we would find Lagrange's equation for generalized coordinate q_2 to be

$$\frac{d}{dt}\left(\frac{\partial T}{\partial \dot{q}_2}\right) - \frac{\partial T}{\partial q_2} = F_x\frac{\partial x}{\partial q_2} + F_y\frac{\partial y}{\partial q_2} + F_z\frac{\partial z}{\partial q_2}. \tag{3.35}$$

3.3 Generalized Forces

We define the right hand side of Lagrange's equation for generalized coordinate q_r to be the *generalized force*, Q_{q_r}, where

$$Q_{q_r} = F_x\frac{\partial x}{\partial q_r} + F_y\frac{\partial y}{\partial q_r} + F_z\frac{\partial z}{\partial q_r}. \tag{3.36}$$

Lagrange's equation for the generalized coordinate q_r can therefore be written as

$$\frac{d}{dt}\left(\frac{\partial T}{\partial \dot{q}_r}\right) - \frac{\partial T}{\partial q_r} = Q_{q_r}. \tag{3.37}$$

We go on later to show that some common forces can be included in Lagrange's equation through the use of *potential energy* but there will often be applied forces that can only enter the equations of motion as generalized forces. These forces are the most difficult for the analyst to handle. In this section, the two most common alternative methods for finding the generalized forces are introduced.

1. *The formal method*

   The most formal approach, and one that always works, starts with the vector expression of the absolute position of the point of application of the force

   $$\vec{p}_F = x\vec{i} + y\vec{j} + z\vec{k}$$

   and then recognizes that the generalized force in Equation 3.36 can be written as

   $$Q_{q_r} = \vec{F} \cdot \frac{\partial \vec{p}_F}{\partial q_r}. \tag{3.38}$$

   Equation 3.38 can be written for each of N applied forces and the resulting scalar generalized forces can be added together to give the total generalized force for generalized coordinate q_r as

   $$Q_{q_r} = \sum_{i=1}^{N} \vec{F}_i \cdot \frac{\partial \vec{p}_{F_i}}{\partial q_r}. \tag{3.39}$$

2. *The intuitive approach*

   Let there be n generalized coordinates specifying the position of a force acting on a dynamic system in Cartesian coordinates. The force will be acting at the point (x, y, z) where the coordinates x, y, and z are functions of the generalized coordinates q_1 through q_n and of time, t, as follows

$$x = x(q_1, q_2, \ldots, q_r, \ldots, q_n, t)$$
$$y = y(q_1, q_2, \ldots, q_r, \ldots, q_n, t)$$
$$z = z(q_1, q_2, \ldots, q_r, \ldots, q_n, t). \tag{3.40}$$

Variations in the position of the force as the generalized coordinates are varied while time is held constant can be written as

$$\delta x = \frac{\partial x}{\partial q_1} \delta q_1 + \frac{\partial x}{\partial q_2} \delta q_2 + \ldots \frac{\partial x}{\partial q_r} \delta q_r + \ldots \frac{\partial x}{\partial q_n} \delta q_n$$
$$\delta y = \frac{\partial y}{\partial q_1} \delta q_1 + \frac{\partial y}{\partial q_2} \delta q_2 + \ldots \frac{\partial y}{\partial q_r} \delta q_r + \ldots \frac{\partial y}{\partial q_n} \delta q_n$$
$$\delta z = \frac{\partial z}{\partial q_1} \delta q_1 + \frac{\partial z}{\partial q_2} \delta q_2 + \ldots \frac{\partial z}{\partial q_r} \delta q_r + \ldots \frac{\partial z}{\partial q_n} \delta q_n. \tag{3.41}$$

If we are trying to find the generalized force corresponding to only one of the generalized coordinates, say q_r, we rewrite Equation 3.41 with $\delta q_i = 0; i = 1, n; i \neq r$ and $\delta q_r \neq 0$, giving

$$\delta x = \frac{\partial x}{\partial q_r} \delta q_r$$
$$\delta y = \frac{\partial y}{\partial q_r} \delta q_r$$
$$\delta z = \frac{\partial z}{\partial q_r} \delta q_r. \tag{3.42}$$

Now consider Equation 3.36 with each side multiplied by δq_r

$$Q_{q_r} \delta q_r = F_x \frac{\partial x}{\partial q_r} \delta q_r + F_y \frac{\partial y}{\partial q_r} \delta q_r + F_z \frac{\partial z}{\partial q_r} \delta q_r. \tag{3.43}$$

Clearly, the terms from Equation 3.42 can be substituted into the right hand side of Equation 3.43 to yield,

$$Q_{q_r} \delta q_r = F_x \delta x + F_y \delta y + F_z \delta z. \tag{3.44}$$

The right hand side of Equation 3.44 can be seen to be the work done by the applied force as its position varies due to changes in the generalized coordinate q_r while all other generalized coordinates and time are held constant.

Using the intuitive approach to finding generalized forces, the analyst will consider, in sequence, the variation of individual generalized coordinates and will write expressions for the total work done during each variation. The generalized force associated with each generalized coordinate will be the work done during the variation of that coordinate divided by the variation in the coordinate. That is,

$$Q_{q_r} = \delta W_{q_r} / \delta q_r. \tag{3.45}$$

3.4 Generalized Forces as Derivatives of Potential Energy

The concept of potential energy is well known to engineers and scientists so it is unnecessary to define it in detail here. We simply state that, if the forces F_x, F_y, and F_z are *conservative*, then they may be written as partial derivatives of a scalar potential energy function, U. That is

$$F_x = -\frac{\partial U}{\partial x}$$
$$F_y = -\frac{\partial U}{\partial y}$$
$$F_z = -\frac{\partial U}{\partial z}. \tag{3.46}$$

Note that the negative signs in Equation 3.46 indicate that we are interested in the *forces applied to the body by the field*.

Figure 3.2 can be used as an example. In this figure, a mass, m, has been raised a distance x vertically above the datum being used for changes in potential energy[4]. We know that the force of gravity on the body is mg, acting vertically downward. If we define the gravitational potential energy to be $U = mgx$, then $F_x = -\frac{\partial U}{\partial x} = -mg$.

Another example of a generalized force that can be included in Lagrange's equation through the potential energy is the force in a linear spring that obeys Hooke's Law[5], $F = kx$ (Figure 3.3). The potential energy in a spring is well known to be

$$U = \frac{1}{2}kx^2$$

where x is the displacement from the unstretched length of the spring. We can then write

$$F_x = -\frac{\partial U}{\partial x} = -kx.$$

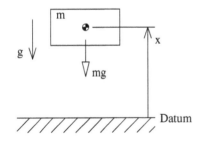

Figure 3.2 Gravitational force acting on a mass.

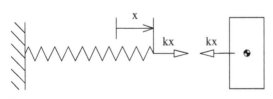

Figure 3.3 Spring force acting on a mass.

---

4 We rarely ask for the "absolute" gravitational potential energy but only for the changes from a reference height ("the datum"), which can be chosen arbitrarily.
5 Robert Hooke (1635–1703), an English scientist, discovered the linear relationship between the force acting on, and the resulting deflection of, elastic bodies.

Substituting Equation 3.46 into Equation 3.36 gives

$$Q_{q_r} = -\frac{\partial U}{\partial x}\frac{\partial x}{\partial q_r} - \frac{\partial U}{\partial y}\frac{\partial y}{\partial q_r} - \frac{\partial U}{\partial z}\frac{\partial z}{\partial q_r} = -\frac{\partial U}{\partial q_r}. \tag{3.47}$$

Therefore, for conservative forces, where the force may be derived from differentiating the potential energy, we may use the result of Equation 3.47. Note that the expression in Equation 3.47 appears on the right hand side of Lagrange's equation. It is usually taken to the left hand side with a change of sign to give

$$\frac{d}{dt}\left(\frac{\partial T}{\partial \dot{q}_r}\right) - \frac{\partial T}{\partial q_r} + \frac{\partial U}{\partial q_r} = Q_{q_r}. \tag{3.48}$$

Lagrange's equation is most often used in the form given in Equation 3.48. That is,

$$\frac{d}{dt}\left(\frac{\partial T}{\partial \dot{q}_r}\right) - \frac{\partial T}{\partial q_r} + \frac{\partial U}{\partial q_r} = Q_{q_r} \tag{3.49}$$

where the potential energy, U, contains all gravitational and elastic effects. All other applied forces are accounted for in the generalized force, Q_{q_r}.

The equation is written once for each of the generalized coordinates chosen to describe the system.

The only assumptions made in the derivation of Lagrange's equation were,

1. The generalized coordinates are independent. That is, each of the generalized coordinates can be arbitrarily varied while all others are held constant. Essentially, this means that the chosen degrees of freedom (i.e. generalized coordinates) should not be related by constraints.
2. The internal constraints are "smooth" and do no work during the variation of a generalized coordinate.

Some authors define a function called the *Lagrangian, L,*

$$L = T - U$$

so that Lagrange's equation may be written as

$$\frac{d}{dt}\left(\frac{\partial L}{\partial \dot{q}_r}\right) - \frac{\partial L}{\partial q_r} = Q_{q_r}$$

where the assumption that potential energy is never a function of velocity has been used. If U actually depended on $\dot{q}_r$, a term $-\frac{d}{dt}\left(\frac{\partial U}{\partial \dot{q}_r}\right)$ would appear.

3.5 Dampers – Rayleigh's Dissipation Function

Devices called "dampers" are common in mechanical systems. These are elements that dissipate energy and they are modeled as producing forces that are proportional to their rate of change of length. The rate of change of length is the relative velocity across the damper. "Proportional" implies linearity and a force proportional to speed implies laminar, viscous flow. As a result, these elements are often referred to as "linear viscous dampers".

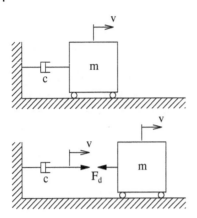

Figure 3.4 A linear viscous damper.

Figure 3.4 shows a system where a body is attached to ground by a damper. The body is moving to the right with speed v and the damping coefficient (constant of proportionality) is c. The physical connection of the damper to both the ground and the body dictates that the rate of change of length of the damper is equal to the speed v. The force in the damper will therefore be $F_d = cv$. The direction of the force will be such that it causes the damper to increase in length as shown in the lower part of Figure 3.4. By Newton's third law, the force on the body must be equal and opposite to the force acting on the damper. The force F_d therefore acts to the left on the body. In other words, the damping force opposes the velocity of the body.

Consider now the more general case of a particle as described in Section 3.2 where the velocity of the body is given by Equation 3.31 as

$$\vec{v}_m = \dot{x}\vec{i} + \dot{y}\vec{j} + \dot{z}\vec{k}. \tag{3.50}$$

Given this velocity, the force that the damper applies to the particle will be

$$\vec{F}_d = -c\vec{v}_m = -c\dot{x}\vec{i} - c\dot{y}\vec{j} - c\dot{z}\vec{k}. \tag{3.51}$$

The components of $\vec{F}_d$ can be substituted into Equation 3.36 to get the following expression for the generalized force arising from the damper

$$Q_{q_r}^d = -c\dot{x}\frac{\partial x}{\partial q_r} - c\dot{y}\frac{\partial y}{\partial q_r} - c\dot{z}\frac{\partial z}{\partial q_r} \tag{3.52}$$

where, using the result from Equation 3.22, we can write

$$\frac{\partial x}{\partial q_r} = \frac{\partial \dot{x}}{\partial \dot{q}_r} \; ; \; \frac{\partial y}{\partial q_r} = \frac{\partial \dot{y}}{\partial \dot{q}_r} \; ; \; \frac{\partial z}{\partial q_r} = \frac{\partial \dot{z}}{\partial \dot{q}_r} \tag{3.53}$$

which can be substituted into Equation 3.52 to yield the following expression for the generalized force

$$Q_{q_r}^d = -c\left(\dot{x}\frac{\partial \dot{x}}{\partial \dot{q}_r} + \dot{y}\frac{\partial \dot{y}}{\partial \dot{q}_r} + \dot{z}\frac{\partial \dot{z}}{\partial \dot{q}_r}\right). \tag{3.54}$$

The generalized force, as expressed in Equation 3.54, can be derived from a scalar function called *Rayleigh's dissipation function* which is defined as

$$\Re = \frac{1}{2}\,c(\vec{v}_m \cdot \vec{v}_m) = \frac{1}{2}c(\dot{x}^2 + \dot{y}^2 + \dot{z}^2). \tag{3.55}$$

A simple differentiation with respect to $\dot{q}_r$ yields

$$\frac{\partial \mathfrak{R}}{\partial \dot{q}_r} = c\left(\dot{x}\frac{\partial \dot{x}}{\partial \dot{q}_r} + \dot{y}\frac{\partial \dot{y}}{\partial \dot{q}_r} + \dot{z}\frac{\partial \dot{z}}{\partial \dot{q}_r}\right) = -Q_{q_r}^d. \tag{3.56}$$

Lagrange's equation can then be written as

$$\frac{\mathrm{d}}{\mathrm{d}t}\left(\frac{\partial T}{\partial \dot{q}_r}\right) - \frac{\partial T}{\partial q_r} + \frac{\partial U}{\partial q_r} = -\frac{\partial \mathfrak{R}}{\partial \dot{q}_r} + Q_{q_r} \tag{3.57}$$

where Q_{q_r} now represents the generalized force corresponding to all externally applied forces that are neither conservative nor linear viscous in nature. Finally, we can transfer the Rayleigh dissipation term to the left hand side and write Lagrange's equation with dissipation as

$$\frac{\mathrm{d}}{\mathrm{d}t}\left(\frac{\partial T}{\partial \dot{q}_r}\right) - \frac{\partial T}{\partial q_r} + \frac{\partial U}{\partial q_r} + \frac{\partial \mathfrak{R}}{\partial \dot{q}_r} = Q_{q_r}. \tag{3.58}$$

3.6 Kinetic Energy of a Free Rigid Body

The extension of Lagrange's equation to rigid bodies follows logically from its application to single particles. We again consider a rigid body to be composed of individual particles that maintain constant inter-particle distances. The rigid body constraint means that each rigid body will have only a limited number of generalized coordinates[6], far fewer than the number of particles in the body.

The analysis requires that the kinetic and potential energy for the rigid body be available. The scalar nature of Lagrange's equations and of the energy of a body allows the total kinetic energy of a rigid body to be found simply by adding the contributions of all the particles.

Figure 3.5 shows the same three-dimensional rigid body that was first introduced in Figure 2.3. The body has a general reference point O with known absolute velocity $\vec{v}_O = v_{O_x}\vec{i} + v_{O_y}\vec{j} + v_{O_z}\vec{k}$. Vectors are expressed in a reference frame having unit vectors $(\vec{i}, \vec{j}, \vec{k})$. The reference frame has angular velocity $\vec{\omega} = \omega_x\vec{i} + \omega_y\vec{j} + \omega_z\vec{k}$. The particle to be considered (particle i) is located with respect to the reference point by the position vector $\vec{p}_{i/O} = x_i\vec{i} + y_i\vec{j} + z_i\vec{k}$ and has mass m_i.

The velocity of particle i can be expressed as

$$\vec{v}_i = \vec{v}_O + \vec{\omega} \times \vec{p}_{i/O}. \tag{3.59}$$

This leads to $\vec{v}_i = v_x\vec{i} + v_y\vec{j} + v_z\vec{k}$, where the components of the velocity of i are

$$v_x = v_{O_x} + \omega_y z_i - \omega_z y_i$$
$$v_y = v_{O_y} + \omega_z x_i - \omega_x z_i$$
$$v_z = v_{O_z} + \omega_x y_i - \omega_y x_i. \tag{3.60}$$

---

6 The maximum is six for a free rigid body that moves without constraint. Three translational coordinates are required to locate the center of mass of the body in space and three angular coordinates are needed to specify its orientation. Any constraint on the motion of the rigid body (e.g. a pin joint) will reduce the number of generalized coordinates.

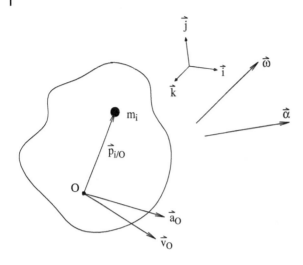

Figure 3.5 A single particle in a rigid body.

The kinetic energy contribution of particle i is then

$$T_i = \frac{1}{2}m_i(\vec{v}_i \cdot \vec{v}_i)$$
$$= \frac{1}{2}m_i(v_x^2 + v_y^2 + v_z^2) \tag{3.61}$$

which, after substituting the expressions for the velocity components and reorganizing, can be written as

$$T_i = \frac{1}{2}m_i[(v_{O_x}^2 + v_{O_y}^2 + v_{O_z}^2) + \omega_x^2(y_i^2 + z_i^2) + \omega_y^2(x_i^2 + z_i^2) + \omega_z^2(x_i^2 + y_i^2)$$
$$- 2\omega_x\omega_y x_i y_i - 2\omega_x\omega_z x_i z_i - 2\omega_y\omega_z y_i z_i$$
$$+ 2v_{O_x}(\omega_y z_i - \omega_z y_i) + 2v_{O_y}(\omega_z x_i - \omega_x z_i) + 2v_{O_z}(\omega_x y_i - \omega_y x_i)]. \tag{3.62}$$

Note that the expression $(v_{O_x}^2 + v_{O_y}^2 + v_{O_z}^2)$ can be replaced by v_O^2 where v_O is the magnitude of the velocity of point O (i.e. the *speed* of point O).

The total kinetic energy of the rigid body is then

$$T = \sum_i T_i = \frac{1}{2}v_O^2 \sum_i m_i$$
$$+ \frac{1}{2}\omega_x^2 \sum_i m_i(y_i^2 + z_i^2) + \frac{1}{2}\omega_y^2 \sum_i m_i(x_i^2 + z_i^2) + \frac{1}{2}\omega_z^2 \sum_i m_i(x_i^2 + y_i^2)$$
$$- \omega_x\omega_y \sum_i m_i x_i y_i - \omega_x\omega_z \sum_i m_i x_i z_i - \omega_y\omega_z \sum_i m_i y_i z_i$$
$$+ v_{O_x}\left(\omega_y \sum_i m_i z_i - \omega_z \sum_i m_i y_i\right) + v_{O_y}\left(\omega_z \sum_i m_i x_i - \omega_x \sum_i m_i z_i\right)$$
$$+ v_{O_z}\left(\omega_x \sum_i m_i y_i - \omega_y \sum_i m_i x_i\right). \tag{3.63}$$

The summation terms in Equation 3.63 have already been considered in Section 2.9. They are

1. $\displaystyle\sum_i m_i = M$

2. $\displaystyle\sum_i m_i(y_i^2 + z_i^2) = I_{xx_O}$

3. $\displaystyle\sum_i m_i(x_i^2 + z_i^2) = I_{yy_O}$

4. $\displaystyle\sum_i m_i(x_i^2 + y_i^2) = I_{zz_O}$

5. $\displaystyle\sum_i m_i x_i y_i = I_{xy_O}$

6. $\displaystyle\sum_i m_i x_i z_i = I_{xz_O}$

7. $\displaystyle\sum_i m_i y_i z_i = I_{yz_O}$

8. $\displaystyle\sum_i m_i x_i = M\bar{x}$

9. $\displaystyle\sum_i m_i y_i = M\bar{y}$

10. $\displaystyle\sum_i m_i z_i = M\bar{z}.$

Making these substitutions into Equation 3.63 gives a *general expression for the kinetic energy of a rigid body*, as follows

$$
\begin{aligned}
T &= \frac{1}{2}M v_O^2 \\
&+ \frac{1}{2}[I_{xx_O}\omega_x^2 + I_{yy_O}\omega_y^2 + I_{zz_O}\omega_z^2 - 2I_{xy_O}\omega_x\omega_y - 2I_{xz_O}\omega_x\omega_z - 2I_{yz_O}\omega_y\omega_z] \\
&+ M[v_{O_x}(\omega_y\bar{z} - \omega_z\bar{y}) + v_{O_y}(\omega_z\bar{x} - \omega_x\bar{z}) + v_{O_z}(\omega_x\bar{y} - \omega_y\bar{x})].
\end{aligned}
\tag{3.64}
$$

There are cases of Equation 3.64 that deserve special attention because, since they cause many of the terms to become zero, they are used almost exclusively in dynamic analysis. These are:

1. *The case where O is a fixed point.*
   In this case, $v_{O_x} = v_{O_y} = v_{O_z} = 0$, and the kinetic energy becomes

   $$
   T = \frac{1}{2}[I_{xx_O}\omega_x^2 + I_{yy_O}\omega_y^2 + I_{zz_O}\omega_z^2 - 2I_{xy_O}\omega_x\omega_y - 2I_{xz_O}\omega_x\omega_z - 2I_{yz_O}\omega_y\omega_z].
   \tag{3.65}
   $$

2. *The case where O is the center of mass, G.*
   In this case, $\bar{x} = \bar{y} = \bar{z} = 0$ and $\bar{v}_O = \bar{v}_G$. The kinetic energy is

   $$
   \begin{aligned}
   T &= \frac{1}{2}M v_G^2 + \frac{1}{2}[I_{xx_G}\omega_x^2 + I_{yy_G}\omega_y^2 + I_{zz_G}\omega_z^2 \\
   &- 2I_{xy_G}\omega_x\omega_y - 2I_{xz_G}\omega_x\omega_z - 2I_{yz_G}\omega_y\omega_z].
   \end{aligned}
   \tag{3.66}
   $$

While the general expression for the kinetic energy is relatively complex, it can be easily remembered for the case where G is the reference point if we simply change mathematical notation. Rather than writing the scalar components of the translational and angular velocities of the body with their respective unit vectors, we can write the velocities as the following column vectors.

$$\{v_G\} = \begin{Bmatrix} v_{G_x} \\ v_{G_y} \\ v_{G_z} \end{Bmatrix} \tag{3.67}$$

$$\{\omega\} = \begin{Bmatrix} \omega_x \\ \omega_y \\ \omega_z \end{Bmatrix}. \tag{3.68}$$

Then, the kinetic energy becomes simply

$$T = \frac{1}{2}M\{v_G\}^T\{v_G\} + \frac{1}{2}\{\omega\}^T[I_G]\{\omega\} \tag{3.69}$$

where $[I_G]$ is the inertia tensor defined in Equations 2.61 and 2.62.

In the case where O is a fixed point, the kinetic energy is simply

$$T = \frac{1}{2}\{\omega\}^T[I_O]\{\omega\}. \tag{3.70}$$

3.7 A Two Dimensional Example using Lagrange's Equation

Figure 3.6 shows a rigid body moving in a plane. Point B on the body is attached to a fixed point A by a massless rigid rod[7] of constant length, r. The body has mass M and moment of inertia about point B of $I_{zz_B} = I$. The unit vectors shown, $(\vec{i}, \vec{j}, \vec{k})$, are fixed in the body. The center of mass, G, is located relative to B by the position vector $\vec{p}_{G/B} = x_0\vec{i} + y_0\vec{j}$.

The problem is to write the equations of motion of the system in terms of the generalized angular coordinates θ and ϕ, each of which measures an angle relative to a fixed direction in space[8].

3.7.1 The Kinetic Energy

We note that this is a two-dimensional problem and that the only possible angular velocities are due to rotations about the $\vec{k}$ direction. The angular velocity of the coordinate system is therefore

$$\vec{\omega} = \dot{\phi}\vec{k}. \tag{3.71}$$

Since the body does not have a fixed point (A is not on the body and B is moving), we can either find the velocity of the center of mass ($\vec{v}_G$) and the moment of inertia about G (I_{zz_G}) and use Equation 3.69 or we can work with the moving reference point B and use Equation 3.64. We choose to work with B in this case.

---

7 This example will be used later and we will consider cases where the rod is inverted and supports the mass from below.

8 θ and ϕ are in fact generalized coordinates because it is possible to vary θ without changing ϕ and vice versa.

Figure 3.6 2D rigid body example.

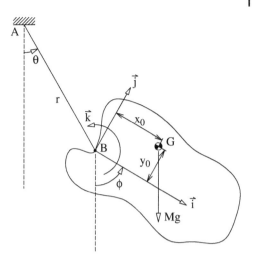

We must first determine the velocity of point B. We start by writing the position of B with respect to A (i.e $\vec{p}_{B/A}$). Since A has no velocity, the absolute velocity of B can be found simply by differentiating $\vec{p}_{B/A}$

$$\vec{p}_{B/A} = r\cos(\phi - \theta)\vec{i} - r\sin(\phi - \theta)\vec{j}. \tag{3.72}$$

The velocity of B is then

$$\vec{v}_B = \frac{d}{dt}\vec{p}_{B/A} = \dot{\vec{p}}_{B/A} + \vec{\omega} \times \vec{p}_{B/A}. \tag{3.73}$$

Consideration of Equation 3.72 shows that there is a rate of change of magnitude of $\vec{p}_{B/A}$ and, since the unit vectors are rotating, there will also be a rate of change of direction. The result, after differentiating and simplifying, is

$$\vec{v}_B = r\dot{\theta}\sin(\phi - \theta)\vec{i} + r\dot{\theta}\cos(\phi - \theta)\vec{j}. \tag{3.74}$$

For this 2D example, Equation 3.64 simplifies to

$$T = \frac{1}{2}Mv_B^2 + \frac{1}{2}I_{zz_B}\dot{\phi}^2 - Mv_{B_x}\dot{\phi}y_0 + Mv_{B_y}\dot{\phi}x_0. \tag{3.75}$$

Substituting for the velocity terms, setting $I_{zz_B} = I$, and simplifying yields

$$T = \frac{1}{2}Mr^2\dot{\theta}^2 + \frac{1}{2}I\dot{\phi}^2 + Mr\dot{\theta}\dot{\phi}[x_0\cos(\phi - \theta) - y_0\sin(\phi - \theta)]. \tag{3.76}$$

3.7.2 The Potential Energy

The next step is to find the potential energy of the system. In this example, there are no springs so all potential energy arises from gravitational effects. Since we are only interested in changes in potential energy from some arbitrary datum, we are free to choose the datum to be at any convenient height. Figure 3.7 shows a suitable datum position corresponding to $\theta = \phi = 0$. As the system moves and θ and ϕ assume non-zero values, the height of the center of mass changes as shown in the figure. The vertical distance from A to G in the two configurations can be used to calculate h, the increase in height of G from the datum.

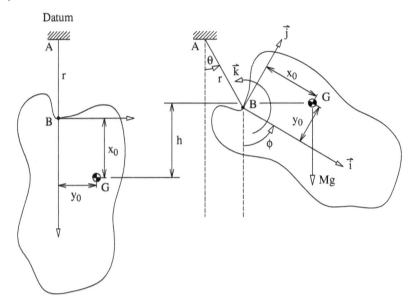

Figure 3.7 2D rigid body potential energy.

The vertical distance from A to G in the datum configuration is $(r + x_0)$. In the displaced configuration, it is $(r \cos \theta + x_0 \cos \phi - y_0 \sin \phi)$. As a result,

$$h = (r + x_0) - (r \cos \theta + x_0 \cos \phi - y_0 \sin \phi)$$

or, after grouping terms,

$$h = r(1 - \cos \theta) + x_0(1 - \cos \phi) + y_0 \sin \phi$$

and, the change in potential energy from the datum to the displaced configuration is

$$U = Mgh = Mg[r(1 - \cos \theta) + x_0(1 - \cos \phi) + y_0 \sin \phi]. \tag{3.77}$$

3.7.3 The θ Equation

We proceed to write each of the terms required in Lagrange's equation for the generalized coordinate θ.

$$\frac{\partial T}{\partial \dot\theta} = Mr^2 \dot\theta - Mr \dot\phi y_0 \sin(\phi - \theta) + Mr \dot\phi x_0 \cos(\phi - \theta) \tag{3.78}$$

$$\frac{d}{dt}\left(\frac{\partial T}{\partial \dot\theta}\right) = Mr^2 \ddot\theta - Mr \ddot\phi y_0 \sin(\phi - \theta) - Mr \dot\phi (\dot\phi - \dot\theta) y_0 \cos(\phi - \theta)$$
$$- Mr \dot\phi (\dot\phi - \dot\theta) x_0 \sin(\phi - \theta) + Mr \ddot\phi x_0 \cos(\phi - \theta) \tag{3.79}$$

$$\frac{\partial T}{\partial \theta} = Mr \dot\theta \dot\phi y_0 \cos(\phi - \theta) + Mr \dot\theta \dot\phi x_0 \sin(\phi - \theta) \tag{3.80}$$

$$\frac{\partial U}{\partial \theta} = Mgr \sin \theta. \tag{3.81}$$

These are then substituted into Equation 3.49 to arrive at the equation of motion for the generalized coordinate θ, which simplifies to

$$Mr^2\ddot{\theta} + Mr\ddot{\phi}[x_0\cos(\phi - \theta) - y_0\sin(\phi - \theta)]$$
$$- Mr\dot{\phi}^2[y_0\cos(\phi - \theta) + x_0\sin(\phi - \theta)] + Mgr\sin\theta = 0. \quad (3.82)$$

3.7.4 The ϕ Equation

The required terms for the generalized coordinate ϕ are

$$\frac{\partial T}{\partial\dot{\phi}} = I\dot{\phi} - Mr\dot{\theta}y_0\sin(\phi - \theta) + Mr\dot{\theta}x_0\cos(\phi - \theta) \quad (3.83)$$

$$\frac{d}{dt}\left(\frac{\partial T}{\partial\dot{\phi}}\right) = I\ddot{\phi} - Mr\ddot{\theta}y_0\sin(\phi - \theta) + Mr\ddot{\theta}x_0\cos(\phi - \theta)$$
$$- Mr\dot{\theta}(\dot{\phi} - \dot{\theta})y_0\cos(\phi - \theta) - Mr\dot{\theta}(\dot{\phi} - \dot{\theta})x_0\sin(\phi - \theta) \quad (3.84)$$

$$\frac{\partial T}{\partial\phi} = -Mr\dot{\theta}\dot{\phi}y_0\cos(\phi - \theta) - Mr\dot{\theta}\dot{\phi}x_0\sin(\phi - \theta) \quad (3.85)$$

$$\frac{\partial U}{\partial\phi} = Mgx_0\sin\phi + Mgy_0\cos\phi. \quad (3.86)$$

The equation of motion for the generalized coordinate ϕ becomes

$$I\ddot{\phi} - Mr\ddot{\theta}[y_0\sin(\phi - \theta) - x_0\cos(\phi - \theta)]$$
$$+ Mr\dot{\theta}^2[y_0\cos(\phi - \theta) + x_0\sin(\phi - \theta)] + Mg[y_0\cos\phi + x_0\sin\phi] = 0. \quad (3.87)$$

3.8 Standard Form of the Equations of Motion

Equations of motion for systems with more than one degree of freedom are often written in a standard form as

$$[M(q, \dot{q}, t)]\{\ddot{q}\} = \{f(q, \dot{q}, t)\} \quad (3.88)$$

where
   $\{q\}$ is the vector of degrees of freedom.
   $\{\dot{q}\}$ and $\{\ddot{q}\}$ are the vectors of first and second derivatives of the degrees of freedom.
   $[M(q, \dot{q}, t)]$ is the *mass matrix* with elements that may depend upon the degrees of freedom, their first derivatives, and time.
   $\{f(q, \dot{q}, t)\}$ is a *forcing vector* containing all of the remaining terms in the equations of motion.

The point to note is that accelerations (i.e. second derivatives of the degrees of freedom) always appear linearly in the equations of motion so that it is always possible to write them in the standard form of Equation 3.88.

Applying this to the equations of motion derived in Section 3.7 yields

$$\begin{bmatrix} Mr^2 & MrP \\ MrP & I \end{bmatrix}\begin{Bmatrix} \ddot{\theta} \\ \ddot{\phi} \end{Bmatrix} = \begin{Bmatrix} Mr\dot{\phi}^2 Q - Mgr\sin\theta \\ -Mr\dot{\theta}^2 Q - Mg[y_0\cos\phi + x_0\sin\phi] \end{Bmatrix} \quad (3.89)$$

where

$$P = x_0 \cos(\phi - \theta) - y_0 \sin(\phi - \theta)$$

and

$$Q = y_0 \cos(\phi - \theta) + x_0 \sin(\phi - \theta).$$

Once the equations of motion have been formulated, the analyst must decide what to do with them. That is, the equations have been derived in order to extract some information about the system being studied. Methods of extracting the required information will be considered in the following chapters. Discussions of equilibrium solutions, stability and mode shapes, frequency response, and time response are included.

Exercises

Descriptions of the systems referred to in the exercises are contained in Appendix A.

3.1 Show that the kinetic energy and potential energy expressions for system 1 are

$$T = \frac{1}{2}(m_1 + m_2)d_1^2\dot{\theta}_1^2 + \frac{1}{2}m_2 d_2^2(\dot{\theta}_1 + \dot{\theta}_2)^2 + m_2 d_1 d_2 \dot{\theta}_1(\dot{\theta}_1 + \dot{\theta}_2)\cos\theta_2$$

and

$$U = (m_1 + m_2)gd_1 \sin\theta_1 + m_2 gd_2 \sin(\theta_1 + \theta_2).$$

3.2 Use Lagrange's equation to show that the equations of motion for system 1 are

$$\begin{bmatrix} (m_1 + m_2)d_1^2 + m_2 d_2(d_2 + 2d_1 \cos\theta_2) & m_2 d_2(d_2 + d_1 \cos\theta_2) \\ m_2 d_2(d_2 + d_1 \cos\theta_2) & m_2 d_2^2 \end{bmatrix}\begin{Bmatrix} \ddot{\theta}_1 \\ \ddot{\theta}_2 \end{Bmatrix}$$
$$= \begin{Bmatrix} m_2 d_1 d_2 \dot{\theta}_2(2\dot{\theta}_1 + \dot{\theta}_2)\sin\theta_2 - (m_1 + m_2)gd_1 \cos\theta_1 - m_2 gd_2 \cos(\theta_1 + \theta_2) \\ -m_2 d_1 d_2 \dot{\theta}_1^2 \sin\theta_2 - m_2 gd_2 \cos(\theta_1 + \theta_2) \end{Bmatrix}.$$

Show that these equations are equivalent to those found in Exercise 2.1.

3.3 Use Lagrange's equation to show that the equations of motion for system 5 are

$$\begin{bmatrix} \frac{3}{2}m & \frac{3}{2}mr \\ \frac{3}{2}mr & \frac{9}{2}mr^2 + m(d-x)^2 \end{bmatrix}\begin{Bmatrix} \ddot{x} \\ \ddot{\theta} \end{Bmatrix}$$
$$= \begin{Bmatrix} mg \sin\theta - m(d-x)\dot{\theta}^2 \\ Fd \cos\theta + mgr \sin\theta - mg(d-x)\cos\theta + 2m(d-x)\dot{x}\dot{\theta} \end{Bmatrix}.$$

Show that these equations are equivalent to those found in Exercise 2.4.

3.4 Use Lagrange's equation to show that the equation of motion for system 6 is

$$mr^2\ddot{\theta} - m\omega^2 r(d - r\cos\theta)\sin\theta + mgr \cos\theta = 0.$$

3.5 Use Lagrange's equation to show that the equation of motion for system 7 is

$$2mL^2 \left(\frac{4}{3} + 12\sin^2\theta \right) \ddot{\theta} + 24mL^2\dot{\theta}^2 \sin\theta \cos\theta + 4kL^2 \sin\theta \cos\theta = 4FL \sin\theta.$$

3.6 Use Lagrange's equation to show that the equation of motion for system 9 is

$$\frac{1}{3}mR^2\ddot{\theta} + kh^2 \sin\theta \cos\theta = FR \sin\theta.$$

3.7 Use Lagrange's equation to show that the equations of motion for system 10 are

$$\begin{bmatrix} m & 0 \\ 0 & md^2 \end{bmatrix} \left\{ \begin{matrix} \ddot{x} \\ \ddot{\theta} \end{matrix} \right\} = \left\{ \begin{matrix} -2kx + kd\sin\theta \\ dF(t) + kxd\cos\theta - kd^2 \sin\theta \cos\theta \end{matrix} \right\}.$$

3.8 Use Lagrange's equation to show that the equation of motion for system 14 is

$$\frac{11}{6}m(R+r)^2\ddot{\theta} - \frac{3}{2}mg(R+r)\sin\theta = -F(R+r)\sin\theta.$$

3.9 Use Lagrange's equation to show that the equations of motion for system 16 are

$$\begin{bmatrix} \frac{4}{3}m\ell^2 + mr^2 + 2mr\ell\cos\phi & \frac{4}{3}m\ell^2 + mr\ell\cos\phi \\ \frac{4}{3}m\ell^2 + mr\ell\cos\phi & \frac{4}{3}m\ell^2 \end{bmatrix} \left\{ \begin{matrix} \ddot{\theta} \\ \ddot{\phi} \end{matrix} \right\}$$

$$= \left\{ \begin{matrix} 2mr\ell\dot{\theta}\dot{\phi}\sin\phi + mr\ell\dot{\phi}^2 \sin\phi + mgr\cos\theta + mg\ell\cos(\theta+\phi) \\ -mr\ell\dot{\theta}^2 \sin\phi + mg\ell\cos(\theta+\phi) \end{matrix} \right\}.$$

3.10 Use Lagrange's equation to show that the equation of motion for system 17 is

$$mr^2\ddot{\theta} - m\omega^2 r^2 \sin\theta \cos\theta = 0.$$

Verify this result using Newton's laws.

3.11 Use Lagrange's equation to derive the equations of motion for system 23 and show that they are equivalent to the equations of motion found in Exercise 2.11.

3.12 Use Lagrange's equation to confirm (see Exercise 2.12) that the equation-of-motion of the rod in system 12, using the angle θ as the single degree-of-freedom, is

$$\frac{4}{3}md^2\ddot{\theta} + 4kd^2 \sin\theta \cos\theta - mgd\cos\theta = 0.$$

Part II

Simulation: Using the Equations of Motion

4

Equilibrium Solutions

Equilibrium solutions of the equations of motion are those where the degrees of freedom assume values that cause their first and second derivatives to go to zero. Under these conditions, there will be no tendency for the values of the degrees of freedom to change and the system will be in an *equilibrium state.*

We will make extensive use of the equilibrium solutions when we consider frequency response and stability of motions.

4.1 The Simple Pendulum

Consider the simple pendulum[1] shown in Figure 4.1. Using the angle θ as the single degree of freedom, the equation of motion is,

$$m\ell^2\ddot{\theta} + mg\ell \sin \theta = 0. \tag{4.1}$$

Once started in motion the pendulum will swing about the point of connection to the ground. In the case of the simple pendulum there is no mechanism for removing energy from the system as it swings (i.e. no friction or other forces that do work) so the motion, once started, will persist.

The motion will depend on the way in which it is started. That is, if the pendulum is rotated to some arbitrary starting angle, θ_0, and released from rest, it will swing through the position where $\theta = 0$ and will eventually return to where it started before reversing and starting the cyclic motion over again. If the pendulum is stopped and returned to θ_0 and then released, not from rest but with an initial velocity, the resulting motion will be different and the pendulum will pass through θ_0 when it returns. The motion will still, however, be cyclic.

Solutions of the equation of motion (Equation 4.1) will predict the results of releasing the pendulum from any initial angle θ_0 with any initial angular velocity. These solutions we leave for later. The question we ask now is *are there initial values of θ where the pendulum can be released from rest and remain stationary?* These are the equilibrium states.

Consider Equation 4.1 under the conditions that there is an initial angle θ_0 and there is no angular velocity (i.e. $\dot{\theta} = 0$ so that θ_0 does not change with time) and that there is no angular

1 The definition of the, often quoted, *simple pendulum* is that it has a massless rigid rod supporting a point mass. The rod is free to swing in a plane about the frictionless point where it is connected to the ground. The only external force is that due to gravity.

The Practice of Engineering Dynamics, First Edition. Ronald J. Anderson.
© 2020 John Wiley & Sons Ltd. Published 2020 by John Wiley & Sons Ltd.
Companion Website: www.wiley.com/go/anderson/engineeringdynamics

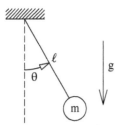

Figure 4.1 A simple pendulum.

acceleration (i.e. $\ddot{\theta} = 0$ so that $\dot{\theta}$ does not change with time and thus there will never be a change in θ_0). This is an equilibrium position and Equation 4.1 becomes,

$$mg\ell \sin\theta_0 = 0. \tag{4.2}$$

Since m and g are never zero, this can only be satisfied by,

$$\sin\theta_0 = 0$$

and the total range of θ is $0 \le \theta \le 2\pi$. In this range, only $\theta = 0$ (the pendulum hangs vertically downward) and $\theta = \pi$ (the pendulum stands upright) satisfy the requirements. These are the two equilibrium states for the pendulum. We will see later that one equilibrium state is stable and the other is unstable.

4.2 Equilibrium with Two Degrees of Freedom

At the end of Chapter 2, the equations of motion for a two dimensional rigid body with two degrees of freedom were derived. These equations, taken from Equation 3.89, are repeated here as Equation 4.3. The system being analyzed is shown in Figure 3.6.

$$\begin{bmatrix} Mr^2 & MrP \\ MrP & I \end{bmatrix} \begin{Bmatrix} \ddot{\theta} \\ \ddot{\phi} \end{Bmatrix} = \begin{Bmatrix} Mr\dot{\phi}^2 Q - Mgr\sin\theta \\ -Mr\dot{\theta}^2 Q - Mg[y_0\cos\phi + x_0\sin\phi] \end{Bmatrix} \tag{4.3}$$

where,

$$P = x_0\cos(\phi - \theta) - y_0\sin(\phi - \theta)$$

and

$$Q = y_0\cos(\phi - \theta) + x_0\sin(\phi - \theta).$$

Finding the equilibrium solutions for this problem is completely analogous to finding the solutions for the simple pendulum. We assume that θ and ϕ have taken up equilibrium values θ_0 and ϕ_0 respectively and that their first and second derivatives are zero. Under these conditions, Equation 4.3 reduces to,

$$\begin{Bmatrix} 0 \\ 0 \end{Bmatrix} = \begin{Bmatrix} -Mgr\sin\theta_0 \\ -Mg[y_0\cos\phi_0 + x_0\sin\phi_0] \end{Bmatrix}. \tag{4.4}$$

There are four solutions to the set of equations presented as Equation 4.4. First note that the two equations in the set are not *coupled* because the first equation contains only θ_0 and the second contains only ϕ_0. As a result, the two equations can be solved independently[2].

---

2 Equations are *coupled* if more than one of the variables describing the motion appears in the same equation.

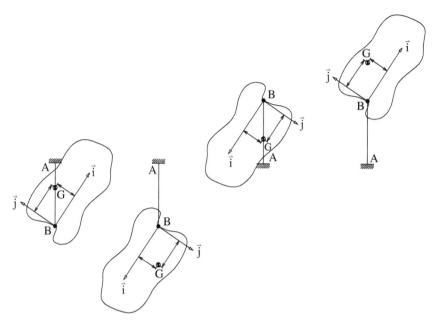

Figure 4.2 The equilibrium solutions for the 2D example.

The first equation is satisfied when $\sin \theta_0 = 0$. The possible range of motion is $0 \leq \theta_0 \leq 2\pi$ and, in this range the only solutions are $\theta_0 = 0$ and $\theta_0 = \pi$.

The second equation is satisfied when

$$\tan \phi_0 = -\frac{y_0}{x_0}. \tag{4.5}$$

There are two angles where this is the case, one in the second quadrant and one in the fourth.

Figure 4.2 shows the four equilibrium solutions for this problem[3]. Notice that all of the equilibrium solutions correspond to a position where the potential energy has an extreme value – either a maximum or a minimum. Notice also that some of the equilibrium solutions are clearly unstable in the sense that the system, if disturbed, will move away from these positions.

4.3 Equilibrium with Steady Motion

Classical static equilibrium is not the only type of equilibrium solution that exists. There are cases where several of the degrees of freedom of a multi-degree of freedom system can have motion that affects the ultimate equilibrium values of the degrees of freedom of interest.

Consider, for example, a system consisting of a large vehicle in which are suspended several objects making up the payload. An analyst may write the equations of motion for the entire system and then decide that the equilibrium values for the degrees of freedom of the payload are of interest in the case where the vehicle is driving in circles at a constant speed.

---

3 Notice the importance of having a massless rigid rod rather than a string in the third and fourth equilibrium positions. The mass cannot stand on a string.

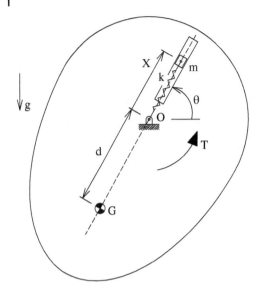

Figure 4.3 An eccentric rotating body.

Another example, perhaps more realistic but again dealing with vehicles, is the case where the *steady-state* value of the roll angle of a car negotiating a long curve at high speed is required as part of the suspension design process. The system is certainly not static but equilibrium values can be found so long as the motion is steady.

Consider, as an example, the system shown in Figure 4.3, which has an eccentric rotating rigid body (mass = M, moment of inertia about the fixed point $O = I$, center of mass at G) in which a slot has been milled to support a small slider (mass= m) that is attached by a spring (stiffness = k, undeflected length = ℓ_0) to the fixed point O. The system has two degrees of freedom, x and θ. There is an applied torque T acting on the rigid body.

The equations of motion for this system are,

$$\begin{bmatrix} I + mx^2 & 0 \\ 0 & m \end{bmatrix} \left\{ \begin{array}{c} \ddot{\theta} \\ \ddot{x} \end{array} \right\} = \left\{ \begin{array}{c} T - 2mx\dot{x}\dot{\theta} + Mgd\cos\theta - mgx\cos\theta \\ mx\dot{\theta}^2 - mg\sin\theta - k(x - \ell_0) \end{array} \right\}. \tag{4.6}$$

We consider two scenarios for establishing equilibrium. The first is in the absence of the applied torque T and the second is in the case where T is specified through a control scheme that maintains a constant angular velocity $\dot{\theta} = \dot{\theta}_0$.

For $T = 0$, we let $x = x_0$, $\dot{x} = \ddot{x} = 0$, $\theta = \theta_0$, and $\dot{\theta} = \ddot{\theta} = 0$ to get the two equilibrium equations,

$$Mgd\cos\theta_0 - mgx_0\cos\theta_0 = 0 \tag{4.7}$$

and

$$-mg\sin\theta_0 - k(x_0 - \ell_0) = 0. \tag{4.8}$$

Equation 4.7 can be written as,

$$(Mgd - mgx_0)\cos\theta_0 = 0 \tag{4.9}$$

which yields three possible equilibrium states.

First, $(Mgd - mgx_0) = 0$ satisfies Equation 4.9. This requires that $x_0 = Md/m$ and then, from Equation 4.8, θ_0 must take the value,

$$\theta_0 = \sin^{-1}\left[\frac{k}{mg}\left(\ell_0 - \frac{Md}{m}\right)\right]. \tag{4.10}$$

What this means physically is that there is a state where the center of mass of the system is at the fixed point O and the system is therefore balanced in an equilibrium position.

If, on the other hand, $(Mgd - mgx_0) \neq 0$, then Equation 4.9 requires that $\cos\theta_0 = 0$ so that $\theta_0 = \pi/2$ and $3\pi/2$ become the equilibrium states. Equation 4.8 then yields $x_0 = \ell_0 - mg/k$ and $x_0 = \ell_0 + mg/k$ for $\theta_0 = \pi/2$ and $3\pi/2$ respectively. These are the equilibrium states where the slot is vertically oriented and the small mass and the center of mass of the rigid body are located on the same vertical line.

We can now consider the case where the torque T is not zero but is, in fact, controlled so that $\dot{\theta} = \dot{\theta}_0$, a constant angular velocity. In this case, the two differential equations making up Equation 4.6 are treated differently. One is used to specify the torque required and the other is used to calculate the equilibrium value of x.

As before, we specify that $x = x_0$, $\dot{x} = \ddot{x} = 0$. The specification of θ is different this time. We define the constant angular velocity $\dot{\theta} = \dot{\theta}_0$ from which we get $\ddot{\theta} = 0$ and $\theta = \dot{\theta}_0 t$. Notice that the angle θ now grows linearly with time. The result of substituting these values into Equation 4.6 is a specification for the required torque,

$$T = (mx_0 - Md)g\cos\left(\dot{\theta}_0 t\right) \tag{4.11}$$

and the value of x_0,

$$x_0 = \frac{k\ell_0 - mg\sin\left(\dot{\theta}_0 t\right)}{k - m\dot{\theta}_0^2}. \tag{4.12}$$

There is clearly a problem here since Equation 4.12 says that the equilibrium value of x is a time varying function. This shows that not every system can have equilibrium values with steady motion. The problem here is the gravitational force that causes the small mass to move harmonically in the slot as the rigid body rotates and the direction of gravity relative to the slot changes.

The same system operating in a horizontal plane yields a much better result. Simply removing the gravitational effects from Equation 4.6 gives the mathematical model of this system in a horizontal plane. The equilibrium result is that $T = 0$, which is to be expected since the system does not have an angular acceleration and,

$$x_0 = \frac{k\ell_0}{k - m\dot{\theta}_0^2} \tag{4.13}$$

indicating that the slider takes up a position where the force in the spring balances the effect of the centripetal acceleration.

It is also possible to have equilibrium states where some degrees of freedom have constant acceleration. Consider for example the system shown in Figure 4.4 where the large body is subjected to the applied force F and the smaller body rolls on top of the large body and is restrained by the spring of stiffness k. We can use the degrees of freedom x and y where x is an absolute coordinate locating the large body relative to ground and y is a relative coordinate locating the small body relative to the large body.

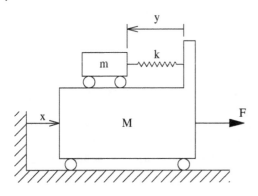

Figure 4.4 The constant acceleration case.

Assuming an undeflected length of ℓ_0 for the spring, the equations of motion are,

$$\begin{bmatrix} (M+m) & -m \\ -m & m \end{bmatrix} \begin{Bmatrix} \ddot{x} \\ \ddot{y} \end{Bmatrix} = \begin{Bmatrix} F \\ -k(y-\ell_0) \end{Bmatrix}. \tag{4.14}$$

It is reasonable to seek the equilibrium value of y given that the force F is controlled so that the large mass has a constant acceleration. That is, we can set $y = y_0 = $ a constant so that $\dot{y} = \ddot{y} = 0$ and let $\ddot{x} = \ddot{x}_0 = $ a constant. Note that this means that x and $\dot{x}$ will both be functions of time. In fact, since $\ddot{x}$ is a constant, we can easily write that $\dot{x} = \ddot{x}_0 t$ and $x = \ddot{x}_0 t^2/2$. These time varying values do not prevent us from finding the equilibrium value of y because neither x nor $\dot{x}$ are present in Equation 4.14.

We therefore use the first equation in 4.14 to find the force required for a constant acceleration,

$$F = (M+m)\ddot{x}_0 \tag{4.15}$$

and the second equation to find,

$$y_0 = \ell_0 + \frac{m\ddot{x}_0}{k}. \tag{4.16}$$

4.4 The General Equilibrium Solution

The standard form of the equations of motion was previously presented as Equation 3.88, which is repeated here for easy reference.

$$[M(q,\dot{q},t)]\{\ddot{q}\} = \{f(q,\dot{q},t)\}. \tag{4.17}$$

The equilibrium solution is simply a set of constant values of the elements of vector $\{q\}$, say $\{q_0\}$, that will satisfy Equation 4.17. Since we are seeking equilibrium values that do not vary with time, the explicit dependence on t disappears from the equations of motion.

Under these conditions, the equations of motion become a set of algebraic equations as follows.

$$\{f(q_0)\} = \{0\}. \tag{4.18}$$

Equation 4.18 will normally contain nonlinear algebraic equations, often requiring numerical solutions.

In some cases, equilibrium states exist under conditions of steady motion. In these cases, a subset of degrees of freedom in the $\{q\}$ vector can be specified as having constant first or second derivatives so long as no time varying elements appear in the equations of motion as a result. If they appear, there will not be an equilibrium solution. The equations of motion corresponding to these degrees of freedom can be used to calculate control inputs or may simply be removed from the analysis. We may therefore say that for the case of steady motion, $\{\dot{q}_0\}$ and $\{\ddot{q}_0\}$ may have some specified, constant, non-zero values and the equilibrium problem can be specified as,

$$\{f(q_0, \dot{q}_0, \ddot{q}_0)\} = \{0\}. \tag{4.19}$$

Exercises

Descriptions of the systems referred to in the exercises are contained in Appendix A.

4.1 Find four equilibrium positions for system 1.

4.2 Show that the equilibrium distance from O to the mass in system 4 is

$$x_0 = \frac{k\,\ell_0}{k - m\,\Omega^2}.$$

4.3 Consider system 6. For the case where the angular velocity is constant at 60 RPM and the disk has a radius of 30 cm, what rod length will result in an equilibrium value of θ equal to 60°? What is the other equilibrium value of θ under these conditions? Note that finding the second equilibrium value will require the use of numerical methods.

4.4 For the case where $F = kL/2$ in system 7, find the two equilibrium values of θ.

4.5 Given that system 8 has reached a steady-state where ω is constant, find the equilibrium positions of the mass attached to the pin A relative to the rotating body. Draw sketches of the system with the mass in its equilibrium positions.

4.6 For system 9, using θ as the single degree-of-freedom, find two equilibrium positions of the rod in the range $0 \le \theta \le \pi/2$.

4.7 Derive the equation governing equilibrium in system 11 and show that setting $mg/kL = 0.07735$ will make the equilibrium value of θ equal to 30°.

4.8 Find the four equilibrium values of θ in system 17.

4.9 For the case where $mg = 2kd$, find four equilibrium positions for the rod in system 12. Sketch the rod in each equilibrium position.

5

Stability

The question of stability of dynamic systems is conceptually simple. In fact, the question is not about the stability of the system as such but about the stability of equilibrium solutions to the equations of motion. The question is: *if the system has reached an equilibrium operating point*[1] *and it is disturbed so that it moves slightly away from the equilibrium state, will it return or not?*

We say the operating point is *stable* if the system returns to it and *unstable* if it tends to move away from it after the disturbance. The information about which way the system will move resides in the equations of motion.

Sometimes stability characteristics become obvious simply by looking at the system as it takes up each of its equilibrium states. A good example of this is the two degree of freedom system presented in Chapter 4 for which the four equilibrium states are shown in Figure 4.2. Clearly, only one equilibrium operating point is stable for this system. The second solution shown has the center of mass at its lowest possible point. If the system moves away from this position, it has no choice but to come back. This is the stable equilibrium operating point for this system.

The three other equilibrium operating points have something in common. In each of these, the center of mass is located at a point where, with a small disturbance, it can be pulled further down by the gravitational force. These three equilibrium operating points are all unstable. We will say more about this later.

5.1 Analytical Stability

Stability analysis is complicated by the nonlinear form of the equations of motion. Everything is simpler if it is linear.

There are two ways of analyzing the stability of a dynamic system about an equilibrium operating point.

1. We may work with the fully nonlinear equations of motion. This leads to analytical techniques that, after a great deal of (interesting) mathematics, give regions in state-space near the equilibrium state where perturbations away from equilibrium will eventually be reduced to zero and the system will return to the equilibrium state if it is stable.

---

1 An operating point is a condition where we know the values of all the degrees of freedom.

The Practice of Engineering Dynamics, First Edition. Ronald J. Anderson.
© 2020 John Wiley & Sons Ltd. Published 2020 by John Wiley & Sons Ltd.
Companion Website: www.wiley.com/go/anderson/engineeringdynamics

This type of analysis is not commonly used for practical problems in mechanics. It is therefore not presented in this work. Readers who are interested in learning about stability of nonlinear equations of motion should start by looking up Lyapunov functions.

2. We may consider a *linearized* version of the equations of motion for small perturbations away from the equilibrium operating point. In this case, the equations of motion will reduce to a set of homogeneous, linear, differential equations for which solutions may be readily written. The solutions tell us if the perturbations away from equilibrium grow or decrease with time, thus establishing the stability of the system about the operating point.

Consider the process of linearizing the equations of motion about an equilibrium operating point. The equations of motion have been derived and written in the form,

$$[M(q, \dot{q}, t)]\{\ddot{q}\} = \{f(q, \dot{q}, t)\}. \tag{5.1}$$

We know from Chapter 4 that the operating point corresponds to an equilibrium solution of Equation 5.1 where $\{q\} = \{q_0\}$, $\{\dot{q}\} = \{\dot{q}_0\}$, $\{\ddot{q}\} = 0$, and the variation with time, t, disappears, giving rise to the equation,

$$\{f(q_0, \dot{q}_0)\} = \{0\}. \tag{5.2}$$

We now consider an infinitesimally small variation, $\{x\}$, away from the equilibrium solution, $\{q_0\}$. The degrees of freedom and their derivatives become,

$$\{q\} \rightarrow \{q_0\} + \{x\}$$
$$\{\dot{q}\} \rightarrow \{\dot{q}_0\} + \{\dot{x}\}$$
$$\{\ddot{q}\} \rightarrow \{\ddot{x}\}. \tag{5.3}$$

Making these substitutions into Equation 5.1 yields,

$$[M(q_0, \dot{q}_0)]\{\ddot{x}\} = \{f(q_0, \dot{q}_0)\} + \{f(x, \dot{x})\}. \tag{5.4}$$

The expression on the left hand side of Equation 5.4 includes a constant matrix, $[M(q_0, \dot{q}_0)]$, which we term the *mass matrix* and represent simply as $[M]$.

The right hand side of Equation 5.4 has two terms, one of which, $\{f(q_0, \dot{q}_0)\}$, is zero by virtue of Equation 5.2. The second term is a vector of nonlinear functions of the infinitesimally small variables, x and $\dot{x}$, which were introduced as perturbations away from equilibrium.

Given that the magnitudes of the elements of $\{x\}$ and $\{\dot{x}\}$ are very small, then products of elements of $\{x\}$ and $\{\dot{x}\}$ (the nonlinear terms) are extremely small and can be ignored relative to linear terms. The result is that terms involving for instance $x_1 x_2$, $x_{10}\dot{x}_4$, $\dot{x}_3^5$ and so on are eliminated from the equations of motion in favor of linear terms.

As a result, the following approximation can be made.

$$\{f(x, \dot{x})\} \approx -[C]\{\dot{x}\} - [K]\{x\} \tag{5.5}$$

where $[K]$ and $[C]$ are constant coefficient matrices, which are called the *stiffness* and *damping* matrices respectively. They are defined with the negative signs so that, when the approximation is substituted into Equation 5.4, it can be written as,

$$[M]\{\ddot{x}\} = -[C]\{\dot{x}\} - [K]\{x\} \tag{5.6}$$

and, immediately, in the standard form for linear equations of motion,

$$[M]\{\ddot{x}\} + [C]\{\dot{x}\} + [K]\{x\} = \{0\}. \tag{5.7}$$

The coefficient matrices are square and of dimension $N \times N$ where N is the number of degrees of freedom in the system. The vector of degrees of freedom, $\{x\}$, and its derivatives are each of length N.

Equation 5.7 is a very interesting and useful result. It says that very small perturbations of the system degrees of freedom away from an equilibrium operating point are governed by a set of second order, linear, differential equations with constant coefficients.

Solutions to differential equations of this type are easy to generate and will tell us whether the perturbations will grow (the *unstable* case) or decay (the *stable* case) with time. The solutions are always exponential functions of time. Think about a function of time which, when multiplied by a constant and added to other constants multiplied by its first two derivatives, gives a result of zero. This is what the differential equation requires and the only function which has the same form as its first two derivatives is the exponential function.

We therefore assume that the solution to Equation 5.7 looks like,

$$\{x(t)\} = \{X\}e^{\lambda t} \tag{5.8}$$

where $\{X\}$ is a vector of complex[2] but constant amplitudes of the degrees of freedom. λ is also a constant complex number but is a scalar rather than a vector.

Equation 5.8 can be differentiated to give,

$$\{\dot{x}(t)\} = \lambda\{X\}e^{\lambda t}$$
$$\{\ddot{x}(t)\} = \lambda^2\{X\}e^{\lambda t}. \tag{5.9}$$

Substituting Equations 5.8 and 5.9 into Equation 5.7 yields,

$$(\lambda^2[M] + \lambda[C] + [K])\{X\}e^{\lambda t} = \{0\}. \tag{5.10}$$

Equation 5.10 can only be satisfied if the left hand side is identically equal to zero. Since $e^{\lambda t}$ can never be zero for any value of λ, it can be taken out of the equation. For a system with N degrees of freedom, the result is a set of N algebraic equations with $N + 1$ unknowns. The unknowns are the N amplitudes contained in $\{X\}$ and the scalar λ, which appears in the coefficient matrix. A general form of the system of equations is,

$$[A(\lambda)]\{X\} = \{0\} \tag{5.11}$$

where $[A(\lambda)] = (\lambda^2[M] + \lambda[C] + [K])$. This is the standard form of an *eigenvalue* problem where there are characteristic values of the eigenvalue, λ, which make the determinant of $[A(\lambda)]$ equal to zero.

Equation 5.11 is a set of N simultaneous algebraic equations with a zero right hand side. If the determinant of $[A(\lambda)]$ is non-zero, then there will be a well-defined, unique solution to the equations and that solution will be $\{X\} = \{0\}$. In other words, the amplitude of every degree of freedom will be zero and the system will not move.

We therefore seek to find values of λ that make the determinant zero. These are the eigenvalues. For each eigenvalue, we determine the value of the amplitude of each degree of

---

2 *Complex* here does not mean *complicated* but rather means that the amplitudes are in the form of *complex numbers.*

freedom and fill in the $\{X\}$ vector. These will be the *eigenvectors*, each corresponding to a different eigenvalue.

Note that we can never solve for all of the elements of $\{X\}$ uniquely. There are only N equations to work with and, as was mentioned earlier, there are $N + 1$ unknowns. We must solve for λ, after which we can only extract $N - 1$ further pieces of information from the set of equations. We, in fact, choose to solve for the ratio of each of the elements of $\{X\}$ to a reference element. That is, for a chosen reference degree of freedom x_p, we find the $N - 1$ complex values of $(x_j/x_p, j = 1, N, j \neq p)$ where, of course, $(x_p/x_p = 1 + 0i)^3$ by definition. This means that the eigenvector shows only the ratio of the amplitudes of the motions in the system and not their actual values.

If we were to write a polynomial to represent the determinant of $[A(\lambda)]$, it would be as follows

$$\det[A(\lambda)] = a_{2N}\lambda^{2N} + a_{2N-1}\lambda^{2N-1} + \cdots + a_1\lambda + a_0 = 0.$$

There are $2N$ roots to this polynomial and therefore $2N$ eigenvalues. The reason for the polynomial being of order $2N$ while $[A(\lambda)]$ is only $N \times N$ is that λ^2 appears in the coefficient matrix.

We therefore have a set of $2N$ eigenvalues and $2N$ eigenvectors. The entire set can be denoted by $(\lambda_j, \{X\}_j, j = 1, 2N)$.

In general, the eigenvalues are complex numbers of the form $\lambda_j = \sigma_j + i\omega_j$. The time functions assumed (see Equation 5.8) as solutions to the linearized differential equations of motion (i.e. $e^{\lambda t}$) now take on the form,

$$e^{\lambda_j t} = e^{(\sigma_j + i\omega_j)t}$$

$$= e^{\sigma_j t} e^{i\omega_j t}$$

$$= e^{\sigma_j t}(\cos\omega_j t + i\sin\omega_j t) \tag{5.12}$$

where the last line of Equation 5.12 uses Euler's formula, $e^{i\theta} = \cos\theta + i\sin\theta$.

Since we are considering the solution of a set of linear differential equations (Equation 5.7), we can apply *superposition* to say that the total response of the system to an arbitrary set of initial conditions will be a linear combination of all of the *natural solutions*. The natural solutions are those that arise from the eigenvalue analysis just performed. Therefore, the total response is,

$$\{x(t)\} = \sum_{j=1}^{2N} c_j\{X\}_j e^{\lambda_j t}$$

$$= c_1\{X\}_1 e^{\lambda_1 t} + c_2\{X\}_2 e^{\lambda_2 t} + \cdots + c_{2N}\{X\}_{2N} e^{\lambda_{2N} t}. \tag{5.13}$$

The constant coefficients in Equation 5.13 (i.e. the c_j) are derived from the initial conditions of the motion. That is, the initial perturbation from the steady-state solution of the nonlinear equations of motion will determine how much of each natural mode enters the response. Since perturbations tend to be random, it is likely that all of the natural modes will be excited from time to time. Therefore, *from the point of view of stability, we must ensure that all of the natural modes decay with time.*

---

3 The notation used here for complex numbers has $i = \sqrt{-1}$

Whether or not a mode decays, depends upon the real part of the eigenvalue. This can be seen from Equation 5.12, where the real part of the eigenvalue (i.e. σ_j) appears in an exponential function that can decay or grow with time depending upon the sign of σ_j. If $\sigma_j > 0$ the exponential term grows inexorably with time. The magnitude of σ_j is related only to the speed at which the function grows and has no bearing on whether it grows or not. Small positive values of σ_j imply slow growth and large positive values of σ_j imply rapid growth but any positive value of σ_j implies that the system will move away from the steady-state position and that the particular operating point is unstable. Conversely, if $\sigma_j < 0$, perturbations away from the steady-state solutions will decay with time and the system will return to the operating point. *Only if all 2N eigenvalues have negative real parts is the system stable.*

The imaginary parts of the eigenvalues appear in the harmonic terms (i.e. sin and cos) of Equation 5.12 and have no effect on whether or not the functions grows or decays. In fact, the magnitude of the expression $(\cos \omega_j t + i \sin \omega_j t)$ is always 1.0. This term shows whether the response to the perturbation is oscillatory or not. If $\omega_j = 0$, then the response is purely exponential and no oscillations occur. If, on the other hand, $\omega_j \neq 0$, then the response is oscillatory and the frequency of oscillation of the jth mode is given by ω_j.

Figure 5.1 shows schematically the response types related to different values of σ_j and ω_j. It contains a number of names that are commonly used in describing the stability characteristics of a system. The two types of instability, *divergent* and *flutter*, are descriptive of the modes of instability and have their origins in aircraft stability analysis performed early in the 20th century. *Neutrally stable* modes have $\sigma_j = 0$ so that no definite statement can be made about whether they are stable or not. In practice neutrally stable modes are considered undesirable and treated as instabilities. *Undamped oscillations* are simply harmonic functions that go on forever without amplitude modulation. *rigid body modes* correspond to eigenvalues that are identically zero and indicate that the system, having no resistance to motion in a particular mode, will simply behave as an unrestrained rigid body which, as Newton said, "perseveres in its state of rest or of uniform motion". Rigid body modes in a

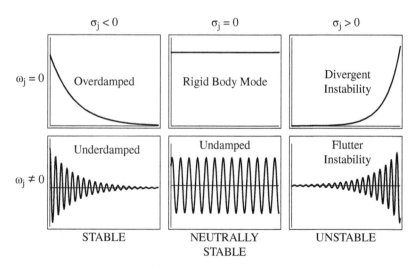

Figure 5.1 Stability response types.

mechanical system often indicate that supporting structures have not been included in the model.

5.2 Linearization of Functions

The linearization of functions about specified operating points is an important topic, deserving of explanation. The trigonometric functions being considered here give us a good opportunity to look closely at the process. We therefore leave the discussion of stability briefly to have a look at linearization.

When asked to give the small angle approximations for the sine and cosine, most engineers will reply that, for small α,

$$\sin \alpha \approx \alpha$$

$$\cos \alpha \approx 1.$$

However, they are unlikely to have given much thought to why this is so or, more importantly, to what assumptions are implicit in these statements.

The small angle approximations are a linearization of the sine and cosine about the operating point $\alpha = 0$. They are, in fact, the first terms arising from a Taylor series[4] expansion of these trigonometric functions about $\alpha = 0$. The expression for a one-dimensional Taylor series expansion for a function $f(x)$ about the point $x = x_0$ is,

$$f(x_0 + \Delta x) = f(x_0) + \frac{df}{dx}\bigg|_{x=x_0} \Delta x + \frac{1}{2!}\frac{d^2f}{dx^2}\bigg|_{x=x_0} \Delta x^2 + \frac{1}{3!}\frac{d^3f}{dx^3}\bigg|_{x=x_0} \Delta x^3 + \cdots . \tag{5.14}$$

If we take x to be α and $f(x) = f(\alpha) = \cos \alpha$, then the derivatives required for the series are,

$$\frac{d(\cos \alpha)}{d\alpha} = -\sin \alpha$$

$$\frac{d^2(\cos \alpha)}{d\alpha^2} = \frac{d(-\sin \alpha)}{d\alpha} = -\cos \alpha$$

$$\frac{d^3(\cos \alpha)}{d\alpha^3} = \frac{d(-\cos \alpha)}{d\alpha} = \sin \alpha$$

and the expansion gives,

$$\cos(\alpha + \Delta\alpha) = \cos \alpha - \Delta\alpha \sin \alpha - \frac{1}{2}\Delta\alpha^2 \cos \alpha + \frac{1}{6}\Delta\alpha^3 \sin \alpha + \cdots . \tag{5.15}$$

To this point the analysis has been general and the relationship just derived can be used to produce an expansion about any angle α. We now consider the expansion about $\alpha = 0$ where $\cos \alpha = 1$ and $\sin \alpha = 0$. The expansion about $\alpha = 0$ therefore yields,

$$\cos \Delta\alpha \approx 1 - \frac{1}{2}\Delta\alpha^2 + \cdots$$

---

4 Brook Taylor (1685–1731), an English mathematician, developed the calculus of finite differences and invented integration by parts. He is best known for introducing the well-known Taylor series.

which, with nonlinear terms ignored because $\Delta\alpha$ is so small that $\Delta\alpha^2$ is negligible[5], gives the common reply by an engineer,

$$\cos \Delta\alpha \approx 1.$$

Doing this again for the *sine* function gives,

$$\frac{d(\sin \alpha)}{d\alpha} = \cos \alpha$$

$$\frac{d^2(\sin \alpha)}{d\alpha^2} = \frac{d(\cos \alpha)}{d\alpha} = -\sin \alpha$$

$$\frac{d^3(\sin \alpha)}{d\alpha^3} = \frac{d(-\sin \alpha)}{d\alpha} = -\cos \alpha$$

from which,

$$\sin(\alpha + \Delta\alpha) = \sin \alpha + \Delta\alpha \cos \alpha - \frac{1}{2}\Delta\alpha^2 \sin \alpha - \frac{1}{6}\Delta\alpha^3 \cos \alpha + \cdots . \tag{5.16}$$

Now let $\alpha = 0$ so that $\cos \alpha = 1$, and $\sin \alpha = 0$. The expansion yields,

$$\sin \Delta\alpha \approx \Delta\alpha - \frac{1}{6}\Delta\alpha^2 + \cdots$$

which, with nonlinear terms ignored, once again yields the common reply,

$$\sin \Delta\alpha \approx \Delta\alpha.$$

Now consider what happens if the expansion is about some angle other than zero. Say that we wish to linearize about the angle $\alpha = \alpha_0$ where $\alpha_0 \neq 0$.
Equation 5.15 becomes,

$$\cos(\alpha_0 + \Delta\alpha) = \cos \alpha_0 - \Delta\alpha \sin \alpha_0 - \frac{1}{2}\Delta\alpha^2 \cos \alpha_0 + \frac{1}{6}\Delta\alpha^3 \sin \alpha_0 + \cdots$$

which can be linearized to,

$$\cos(\alpha_0 + \Delta\alpha) \approx \cos \alpha_0 - \Delta\alpha \sin \alpha_0.$$

Similarly, Equation 5.16 becomes,

$$\sin(\alpha_0 + \Delta\alpha) = \sin \alpha_0 + \Delta\alpha \cos \alpha_0 - \frac{1}{2}\Delta\alpha^2 \sin \alpha_0 - \frac{1}{6}\Delta\alpha^3 \cos \alpha_0 + \cdots .$$

In linear form, this is,

$$\sin(\alpha_0 + \Delta\alpha) \approx \sin \alpha_0 + \Delta\alpha \cos \alpha_0.$$

As an aside, these same relationships can be derived from the well known "sum of angles formulae"[6]

$$\cos(\alpha + \beta) = \cos \alpha \cos \beta - \sin \alpha \sin \beta$$

---

5 The term *linearization* expresses the fact that the perturbation in the variable being considered is so small that squaring it, cubing it, or taking it to any higher power produces terms whose magnitudes are negligibly small compared to the linear term and can therefore be ignored so that retaining the linear term only provides a good approximation to the value of the function very close to the operating point.
6 While these may be "well known", they are not easily remembered. However, they can be quickly derived using Euler's formula: $e^{i\theta} = \cos \theta + i \sin \theta$. We write the equation twice, once for the angle α and again for the angle β, giving,

$$e^{i\alpha} = \cos \alpha + i \sin \alpha$$

$$\sin(\alpha + \beta) = \cos \alpha \sin \beta + \sin \alpha \cos \beta.$$

We substitute α_0 for α and $\Delta\alpha$ for β and get,

$$\cos(\alpha_0 + \Delta\alpha) = \cos \alpha_0 \cos \Delta\alpha - \sin \alpha_0 \sin \Delta\alpha$$

$$\sin(\alpha_0 + \Delta\alpha) = \cos \alpha_0 \sin \Delta\alpha + \sin \alpha_0 \cos \Delta\alpha$$

and then we use the "usual" small angle approximation to say that,

$$\cos \Delta\alpha \approx 1$$

and

$$\sin \Delta\alpha \approx \Delta\alpha.$$

Substituting these above yields,

$$\cos(\alpha_0 + \Delta\alpha) \approx \cos \alpha_0 - \Delta\alpha \sin \alpha_0$$

$$\sin(\alpha_0 + \Delta\alpha) \approx \sin \alpha_0 + \Delta\alpha \cos \alpha_0.$$

These are the same results obtained with the Taylor series expansion.

A trigonometric function of two angles (e.g. $\sin(\alpha + \beta)$ or $\cos(\alpha + \beta)$) is a function of two variables so that the Taylor series expansion of Equation 5.14 is insufficient for the purpose. Instead we use a two variable form of the Taylor series expansion of the function $f(x, y)$ about the operating point (x_0, y_0).

$$f(x_0 + \Delta x, y_0 + \Delta y) = f(x_0, y_0) + \left.\frac{\partial f}{\partial x}\right|_{x_0, y_0} \Delta x + \left.\frac{\partial f}{\partial y}\right|_{x_0, y_0} \Delta y$$

$$+ \frac{1}{2!} \left(\left.\frac{\partial^2 f}{\partial x^2}\right|_{x_0, y_0} \Delta x^2 + \left.\frac{\partial^2 f}{\partial x \partial y}\right|_{x_0, y_0} \Delta x \Delta y + \left.\frac{\partial^2 f}{\partial y^2}\right|_{x_0, y_0} \Delta y^2 \right) + \cdots .$$

$$(5.17)$$

The point to note is that, whether we are using the single variable expansion (Equation 5.14) or the two variable expansion (Equation 5.17), the linear terms involve only the first derivative(s) of the function[7]. Second and higher derivatives always involve

---

$$e^{i\beta} = \cos \beta + i \sin \beta$$

We then multiply these two equations together giving, on the left hand side,

$$e^{i\alpha} e^{i\beta} = e^{i(\alpha + \beta)} = \cos(\alpha + \beta) + i \sin(\alpha + \beta)$$

and, on the right hand side,

$$(\cos \alpha \cos \beta - \sin \alpha \sin \beta) + i(\cos \alpha \sin \beta + \sin \alpha \cos \beta)$$

Equating the real and imaginary parts of these two results gives the "sum of angles" formulae we couldn't remember,

$$\cos(\alpha + \beta) = \cos \alpha \cos \beta - \sin \alpha \sin \beta$$

$$\sin(\alpha + \beta) = \cos \alpha \sin \beta + \sin \alpha \cos \beta.$$

7 The linear terms in multi-variable Taylor series expansions also involve only the first derivatives of the function.

Figure 5.2 Linearization.

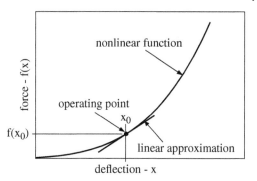

nonlinear terms that will eventually be ignored. Linearization of functions about operating points is therefore an exercise in finding the local slope of the function and assuming that small deviations away from the operating point can be approximated by points lying on this straight line. This is shown graphically in Figure 5.2 for a function of one variable and can be generalized to functions of many variables.

The linearized version of a function of many variables, $f(x, y, z, \ldots)$, about the operating point, $(x_0, y_0, z_0, \ldots)$, is therefore,

$$f(x_0 + \Delta x, y_0 + \Delta y, z_0 + \Delta z, \ldots) = f(x_0, y_0, z_0, \ldots)$$
$$+ \left.\frac{\partial f}{\partial x}\right|_{x_0, y_0, z_0, \ldots} \Delta x + \left.\frac{\partial f}{\partial y}\right|_{x_0, y_0, z_0, \ldots} \Delta y + \left.\frac{\partial f}{\partial z}\right|_{x_0, y_0, z_0, \ldots} \Delta z + \cdots . \tag{5.18}$$

If we apply Equation 5.18 to $\cos(\alpha + \beta)$, for instance, we first need to find,

$$\frac{\partial \cos(\alpha + \beta)}{\partial \alpha} = -\sin(\alpha + \beta)$$

and

$$\frac{\partial \cos(\alpha + \beta)}{\partial \beta} = -\sin(\alpha + \beta)$$

then, substituting into Equation 5.18, we find,

$$\cos[(\alpha + \Delta \alpha) + (\beta + \Delta \beta)] \approx \cos(\alpha + \beta) - \Delta \alpha \sin(\alpha + \beta) - \Delta \beta \sin(\alpha + \beta).$$

Similarly, the linearized version of $\sin(\alpha + \beta)$ is,

$$\sin[(\alpha + \Delta \alpha) + (\beta + \Delta \beta)] \approx \sin(\alpha + \beta) + \Delta \alpha \cos(\alpha + \beta) + \Delta \beta \cos(\alpha + \beta).$$

Equation 5.18 can be used whenever a function needs to be linearized about a known operating point.

5.3 Example: A System with Two Degrees of Freedom

At the end of Chapter 3, the equations of motion for a two dimensional rigid body (see Figure 3.6) with two degrees of freedom were derived. The equilibrium positions of this system were found in Section 4.2 and are demonstrated graphically in Figure 4.2. Here, we consider the stability of each of the four equilibrium positions.

Before proceeding with the analysis we should inspect the equilibrium positions and see how we would expect the system to behave if it were moved very slightly away from equilibrium. Referring to Figure 4.2, it should be clear that the first, third, and fourth equilibrium positions are unstable. Each is comparable to an inverted pendulum that, given a slight disturbance, will fall. Further, we expect that these instabilities will be divergent. That is, there will be no oscillations that cause the system to pass through the equilibrium position many times as the amplitude of motion increases. They will simply fall. The second equilibrium is different because the center of mass of the rigid body is at its lowest possible position and cannot fall. If we disturb this equilibrium we should see a neutrally stable harmonic motion about the equilibrium. The motion is not stable unless the amplitudes decrease with time and, since there is no element in this system which removes energy, we expect undamped oscillations.

The equations of motion for the system, taken from Equation 3.89, are repeated here as Equation 5.19.

$$\begin{bmatrix} Mr^2 & MrP \\ MrP & I \end{bmatrix} \begin{Bmatrix} \ddot{\theta} \\ \ddot{\phi} \end{Bmatrix} = \begin{Bmatrix} Mr\dot{\phi}^2 Q - Mgr\sin\theta \\ -Mr\dot{\theta}^2 Q - Mg[y_0\cos\phi + x_0\sin\phi] \end{Bmatrix} \tag{5.19}$$

where,

$$P = x_0\cos(\phi - \theta) - y_0\sin(\phi - \theta)$$

and

$$Q = y_0\cos(\phi - \theta) + x_0\sin(\phi - \theta).$$

This example provides quite a challenge in terms of linearization. Some terms in the equations of motion are obviously nonlinear and can be discarded immediately. These are the terms on the right hand side containing $\dot{\phi}^2$ and $\dot{\theta}^2$. What makes the linearization difficult are the trigonometric functions of more than one variable such as $\cos(\phi - \theta)$ and $\sin(\phi - \theta)$. *Linearization* was discussed in Section 5.2 and the techniques described there will be used here.

In order to derive the linearized version of Equation 5.19, we first recognize that the terms in the mass matrix must be constant. That is, they cannot be linear functions of any of the degrees of freedom since each of these terms is already multiplied by the second derivatives of the degrees of freedom that reside in the column vector to the right of the mass matrix. The expansion for the $[x_0\cos(\phi - \theta) - y_0\sin(\phi - \theta)]$ term is therefore simply $[x_0\cos(\phi_0 - \theta_0) - y_0\sin(\phi_0 - \theta_0)]$, keeping only the constant terms from the expansions of $\cos(\phi - \theta)$ and $\sin(\phi - \theta)$ as discussed in Section 5.2. During the linearizing process, the equilibrium expressions of Equation 4.4 are also found and are set equal to zero because of the requirement that they be zero at an equilibrium point.

On the right hand side, we immediately drop the terms containing squares of angular velocities as being nonlinear. The remaining expressions involve quite simple linearizations of sine and cosine, which, when done, give the linearized equations of motion as,

$$\begin{bmatrix} Mr^2 & MrP \\ MrP & I \end{bmatrix} \begin{Bmatrix} \Delta\ddot{\theta} \\ \Delta\ddot{\phi} \end{Bmatrix}$$

$$+ \begin{bmatrix} Mgr\cos\theta_0 & 0 \\ 0 & Mg[-y_0\sin\phi_0 + x_0\cos\phi_0] \end{bmatrix} \begin{Bmatrix} \Delta\theta \\ \Delta\phi \end{Bmatrix} = \begin{Bmatrix} 0 \\ 0 \end{Bmatrix} \tag{5.20}$$

where,

$$P = [x_0 \cos(\phi_0 - \theta_0) - y_0 \sin(\phi_0 - \theta_0)].$$

We now assume an exponential solution to the linear differential equation (Equation 5.20) and let,

$$\left\{ \begin{array}{c} \Delta\theta \\ \Delta\phi \end{array} \right\} = \left\{ \begin{array}{c} X_1 \\ X_2 \end{array} \right\} e^{\lambda t}$$

which gives, after two differentiations,

$$\left\{ \begin{array}{c} \Delta\ddot{\theta} \\ \Delta\ddot{\phi} \end{array} \right\} = \lambda^2 \left\{ \begin{array}{c} X_1 \\ X_2 \end{array} \right\} e^{\lambda t}.$$

Substituting into Equation 5.20 gives,

$$\begin{bmatrix} Mr^2\lambda^2 + Mgr\cos\theta_0 & MrP\lambda^2 \\ MrP\lambda^2 & I\lambda^2 + Mg[-y_0\sin\phi_0 + x_0\cos\phi_0] \end{bmatrix} \left\{ \begin{array}{c} X_1 \\ X_2 \end{array} \right\} e^{\lambda t}$$

$$= \left\{ \begin{array}{c} 0 \\ 0 \end{array} \right\}. \qquad (5.21)$$

To find the eigenvalues, we write the determinant of the coefficient matrix in Equation 5.21 as a polynomial in λ and set it equal to zero. The polynomial is,

$$\lambda^4[Mr^2(I - MP^2)] + \lambda^2\{Mgr[I\cos\theta_0 + Mr(-y_0\sin\phi_0 + x_0\cos\phi_0)]\}$$
$$+ M^2g^2r\cos\theta_0(-y_0\sin\phi_0 + x_0\cos\phi_0) = 0. \qquad (5.22)$$

It is clear that, even with this relatively simple example, making general comments about the stability of the system by working algebraically with the polynomial in Equation 5.22 will be difficult. In fact, we will be able to say something about the stability later when we consider a different method. For the time being, we will assign values to the parameters in the system and calculate the eigenvalues numerically to see if they agree with the intuitive results presented at the beginning of this section.

Consider the case where the system has parameter[8] values as specified in Table 5.1. The resulting eigenvalues and eigenvectors of the system are presented in Table 5.2.

The data presented in Table 5.2 describe the natural motions of the system when it is moved slightly away from an equilibrium state. Reference to Figure 4.2 will be helpful in understanding the descriptions that follow.

- Case 1: There is a single unstable[9] mode (i.e. an eigenvalue with a positive real part) in this equilibrium position. The mode shape (i.e. the eigenvector) indicates that ϕ is much greater than θ for this mode[10]. Referring to Figure 3.6 for the definitions of the degrees of freedom, θ and ϕ, and to Figure 4.2 for a view of the system when it is the equilibrium position of case 1, it is clear that the mass will tend to fall and that changes in the angle ϕ will be the dominant motion as is does. This is a divergent instability.

---

8 Equation 4.5 expresses the equilibrium value of ϕ as a function of x_0 and y_0. The parameter values chosen here result in an angle that is approximately $-26.5°$ (-0.46365 rad) or π plus this value, depending on the case. Similarly, the equilibrium values of θ were found to be 0 or π.

9 The eigenvalues related to unstable modes are shown in boldface in Table 5.2.

10 Remember that the eigenvector is defined with θ as the first element and ϕ as the second.

Table 5.1 Parameter values for the two degree of freedom example.

Parameter	Symbol	Value	Units
Mass	M	10.0	kg
Moment of inertia about B	I	5.5	kg m^2
Length of supporting rod	r	0.5	m
Distance from B to G ($\vec{i}$ direction)	x_0	0.2	m
Distance from B to G ($\vec{j}$ direction)	y_0	0.1	m
Acceleration due to gravity	g	9.807	m s^{-2}
Supporting rod angle (case 1)	θ_0	0	rad
Mass angle (case 1)	ϕ_0	π-0.46365	rad
Supporting rod angle (case 2)	θ_0	0	rad
Mass angle (case 2)	ϕ_0	−0.46365	rad
Supporting rod angle (case 3)	θ_0	π	rad
Mass angle (case 3)	ϕ_0	−0.46365	rad
Supporting rod angle (case 4)	θ_0	π	rad
Mass angle (case 4)	ϕ_0	π-0.46365	rad

Table 5.2 Eigenvalues and eigenvectors for the two degree of freedom example.

Case	Eigenvalue/eigenvector			
1	$\lambda_1 = -4.6087i$	$\lambda_2 = 4.6087i$	$\lambda_3 = \mathbf{2.0125}$	$\lambda_4 = -2.0125$
	$\{X\}_1 = \left\{\begin{array}{c} 1.0 \\ 0.1712 \end{array}\right\}$	$\{X\}_2 = \left\{\begin{array}{c} 1.0 \\ 0.1712 \end{array}\right\}$	$\{X\}_3 = \left\{\begin{array}{c} 0.07654 \\ 1.0 \end{array}\right\}$	$\{X\}_4 = \left\{\begin{array}{c} 0.07654 \\ 1.0 \end{array}\right\}$
2	$\lambda_1 = -4.6970i$	$\lambda_2 = 4.6970i$	$\lambda_3 = -1.9746i$	$\lambda_4 = 1.9746i$
	$\{X\}_1 = \left\{\begin{array}{c} 1.0 \\ -0.2481 \end{array}\right\}$	$\{X\}_2 = \left\{\begin{array}{c} 1.0 \\ -0.2481 \end{array}\right\}$	$\{X\}_3 = \left\{\begin{array}{c} 0.1110 \\ 1.0 \end{array}\right\}$	$\{X\}_4 = \left\{\begin{array}{c} 0.1110 \\ 1.0 \end{array}\right\}$
3	$\lambda_1 = \mathbf{4.6087}$	$\lambda_2 = -4.6087$	$\lambda_3 = -2.0125i$	$\lambda_4 = 2.0125i$
	$\{X\}_1 = \left\{\begin{array}{c} 1.0 \\ 0.1712 \end{array}\right\}$	$\{X\}_2 = \left\{\begin{array}{c} 1.0 \\ 0.1712 \end{array}\right\}$	$\{X\}_3 = \left\{\begin{array}{c} 0.07654 \\ 1.0 \end{array}\right\}$	$\{X\}_4 = \left\{\begin{array}{c} 0.07654 \\ 1.0 \end{array}\right\}$
4	$\lambda_1 = \mathbf{4.6970}$	$\lambda_2 = -4.6970$	$\lambda_3 = \mathbf{1.9746}$	$\lambda_4 = -1.9746$
	$\{X\}_1 = \left\{\begin{array}{c} 1.0 \\ -0.2481 \end{array}\right\}$	$\{X\}_2 = \left\{\begin{array}{c} 1.0 \\ -0.2481 \end{array}\right\}$	$\{X\}_3 = \left\{\begin{array}{c} 0.1110 \\ 1.0 \end{array}\right\}$	$\{X\}_4 = \left\{\begin{array}{c} 0.1110 \\ 1.0 \end{array}\right\}$

- Case 2: Case 2 is the only stable equilibrium position for this system. Here, all of the eigen-values are purely imaginary, indicating undamped oscillatory motion. That this state is only neutrally stable is no surprise since there are no forces that remove energy from the system. To be truly stable some "damping" terms would need to be added to the system. There are two distinct modes in case 2. There is an oscillation at 4.6970 rad s^{-1} (approximately 0.75 Hz) where changes in the angle θ are dominant. There is a lower frequency oscillation (1.9746 rad s^{-1} or 0.31 Hz), which is dominated by changes in the angle ϕ.
- Case 3: There is a single divergent instability here which will demonstrate itself as a change in the angle θ. Again, this makes sense when we look at case 3 in Figure 4.2 and see that, in order to "fall", the system will see large changes in θ.
- Case 4: There are two divergent instabilities in this equilibrium state. Again referring to the figure, this makes sense. Both θ and ϕ will see large changes as the system leaves this precarious equilibrium state. The mode shapes of the two instabilities indicate that each is dominated by one of the two angles.

In reality, systems never stay in unstable equilibrium positions. There will always be a slight disturbance that will start the system moving and it will always move from the unstable equilibrium to a stable equilibrium position. In this example, only case 2 was a stable equilibrium and the system would eventually find itself there. Whether an equilibrium state is stable or not is not as obvious for systems more complex than this example.

5.4 Routh Stability Criterion

In the previous section, we derived the polynomial form of the determinant[11] of the coefficient matrix for the linearized equations of motion of the system for small motions away from an equilibrium point (Equation 5.22). We decided it was too difficult to pursue an algebraic solution to the polynomial and proceeded to handle it numerically. We now return to the polynomial and consider its stability using the *Routh*[12] *stability criterion*.

The Routh stability criterion is a method for determining system stability from the nth order characteristic equation of the form,

$$a_n \lambda^n + a_{n-1} \lambda^{n-1} + \cdots + a_1 \lambda + a_0. = 0 \tag{5.23}$$

From the coefficients of the characteristic equation, we form a *Routh table* as shown in Table 5.3. The first two rows of the table contain the coefficients of the characteristic equation and are padded with zeros at the end. We find the coefficients b_l and c_i from 2×2 determinants involving the coefficients in the two rows just preceding them. That is,

$$b_1 = \frac{a_{n-1} a_{n-2} - a_n a_{n-3}}{a_{n-1}} \quad b_2 = \frac{a_{n-1} a_{n-4} - a_n a_{n-5}}{a_{n-1}} \quad \cdots$$

$$\tag{5.24}$$

$$c_1 = \frac{b_1 a_{n-3} - a_{n-1} b_2}{b_1} \quad c_2 = \frac{b_1 a_{n-5} - a_{n-1} b_3}{b_1} \quad \cdots .$$

---

11 This polynomial is called the *characteristic polynomial* for the system
12 Edward John Routh (1831–1907), born in Canada, studied mathematics in England and was well known for the teaching skills he demonstrated while employed at Cambridge University. He is especially remembered for his contributions to the field of dynamic stability.

Table 5.3 The Routh table.

λ^n	a_n	a_{n-2}	a_{n-4}	$\cdots$	0
λ^{n-1}	a_{n-1}	a_{n-3}	a_{n-5}	$\cdots$	0
λ^{n-2}	b_1	b_2	b_3	$\cdots$	
λ^{n-3}	c_1	c_2	c_3	$\cdots$	
$\vdots$	$\vdots$	$\vdots$	$\vdots$	$\vdots$	

Each new element generated for the table depends on the leading elements of the preceding two rows and on the two elements just to the right on the same two rows. For instance, the calculation of c_2, the element in row 4 and column 2, uses a_{n-1} and b_1, the two leading elements in rows 2 and 3, and a_{n-5} and b_3, the two elements in column 3 of rows 2 and 3. The pattern is easy to follow and the table is continued horizontally and vertically until only zeros are obtained.

A statement on the stability of the equilibrium state being analyzed for the system can be made from consideration of the first column of the Routh table as follows: *all of the roots of the characteristic equation have negative real parts if and only if the elements of the first column of the Routh table have the same sign. Otherwise, the number of roots with positive real parts is equal to the number of sign changes in the first column.*

Remember that the roots of the characteristic equation are the eigenvalues of the linearized equations of motion of the system for small perturbations about an equilibrium state. Our previous work said that having all eigenvalues with negative real parts ensured stability. Therefore, having the same sign, whether positive or negative, on all of the elements in the first column of the Routh table ensures stability. If the column starts out being positive, for example, and then changes to a negative value at some element, staying negative thereafter, there is a single sign change and, as a result, one unstable mode. If, on the other hand, there is a change of sign as just described but it is followed later by a return to the original sign, there will be two unstable modes and so on.

We now return to the two degree of freedom system for which we have been considering the stability. In particular, we reconsider getting an analytical statement on stability by reconsidering the characteristic equation first presented as Equation 5.22 and reproduced below as Equation 5.25.

$$\lambda^4[Mr^2(I - MP^2)] + \lambda^2\{Mgr[I \cos\theta_0 + Mr(-y_0 \sin\phi_0 + x_0 \cos\phi_0)]\}$$
$$+ M^2g^2r \cos\theta_0(-y_0 \sin\phi_0 + x_0 \cos\phi_0) = 0 \tag{5.25}$$

where

$$P = [x_0 \cos(\phi_0 - \theta_0) - y_0 \sin(\phi_0 - \theta_0)]. \tag{5.26}$$

The characteristic equation is of the form,

$$a_4\lambda^4 + a_3\lambda^3 + a_2\lambda^2 + a_1\lambda + a_0 = 0 \tag{5.27}$$

where

$$a_4 = Mr^2(I - MP^2)$$

Table 5.4 The initial Routh table for the two degrees of freedom example.

λ^4	a_4	a_2	a_0	0
λ^3	ϵ	0	0	0
λ^2	b_1	b_2	0	
λ^1	c_1	c_2		
λ^0	d_1			

$$a_3 = 0$$
$$a_2 = Mgr[I\cos\theta_0 + Mr(-y_0\sin\phi_0 + x_0\cos\phi_0)]$$
$$a_1 = 0$$
$$a_0 = M^2g^2r\cos\theta_0(-y_0\sin\phi_0 + x_0\cos\phi_0). \tag{5.28}$$

The Routh table for this case starts out as shown in Table 5.4 where a small but non-zero term ϵ has been introduced as the first element in the second row. We work with ϵ to avoid division by zero while we fill in the remainder of the table and then we let $\epsilon \to 0$ to get the final form of the table.

The elements in the table are calculated as follows.

$$b_1 = \frac{\epsilon \cdot a_2 - a_4 \cdot 0}{\epsilon} = a_2$$
$$b_2 = \frac{\epsilon \cdot a_0 - a_4 \cdot 0}{\epsilon} = a_0$$
$$c_1 = \frac{b_1 \cdot 0 - \epsilon \cdot b_2}{b_1} = -\frac{\epsilon \cdot b_2}{b_1}$$
$$c_2 = \frac{b_1 \cdot 0 - \epsilon \cdot 0}{b_1} = 0$$
$$d_1 = \frac{c_1 \cdot b_2 - b_1 \cdot c_2}{c_1} = a_0.$$

Making these substitutions into Table 5.4 and letting $\epsilon \to 0$ yields the final Routh table presented in Table 5.5.

The first column of Table 5.5 contains all of the coefficients in the characteristic equation plus three zeros. The zeros are representative of the same neutrally stable behavior we noted

Table 5.5 The final Routh table for the two degrees of freedom example.

λ^4	a_4	a_2	a_0	0
λ^3	0	0	0	0
λ^2	a_2	a_0	0	
λ^1	0	0		
λ^0	a_0			

earlier. They cannot be taken as changes of sign but rather of a "near" change in sign. What is of more interest to us is that all of the coefficients in the characteristic equation must have the same sign or the system will be unstable.

The coefficients are given in Equation 5.28. It is immediately clear that it will be difficult to make comments about the signs of any of these coefficients if we do not have relatively easy values of θ_0 and ϕ_0 to work with. For that reason, we will consider the case where $y_0 = 0$. This means that the center of gravity G is located a distance x_0 from point B and that the equilibrium values of ϕ (found from Equation 4.5) are,

$$\phi_0 = \arctan(-\frac{y_0}{x_0}) = \arctan(0) = \begin{cases} 0 \\ \pi \end{cases}.$$

The four equilibrium cases, expressed as their values of the angles θ_0 and ϕ_0 and the resulting value of p from Equation 5.26 are now,

- Case 1: $\theta_0 = 0$, $\phi_0 = \pi \Rightarrow P = -x_0$.
- Case 2: $\theta_0 = 0$, $\phi_0 = 0 \Rightarrow P = x_0$.
- Case 3: $\theta_0 = \pi$, $\phi_0 = 0 \Rightarrow P = -x_0$.
- Case 4: $\theta_0 = \pi$, $\phi_0 = \pi \Rightarrow P = x_0$.

We first consider the sign of $a_4 = Mr^2(I - MP^2)$. In all of the equilibrium cases being considered, the value of a_4 is $a_4 = Mr^2(I - Mx_0^2)$. This is, using the *parallel axis theorem*[13], $a_4 = Mr^2 I_G > 0$ since I_G is greater than zero by definition.

Having $a_4 > 0$ sets the stage for the Routh stability calculation. The system will be stable in the chosen equilibrium if and only if $a_2 > 0$ and $a_0 > 0$.

Consider case 1. We have $\theta_0 = 0$ and $\phi_0 = \pi$. As a result, with $y_0 = 0$,

$$a_2 = Mgr[I \cos \theta_0 + Mr(-y_0 \sin \phi_0 + x_0 \cos \phi_0)] = Mgr(I - Mrx_0).$$

The sign of a_2 depends on the relative magnitudes of I and MRx_0. Given the parameter values we are using for this analysis, $I = 5.5$ and $Mrx_0 = 1.0$ so that $a_2 > 0$.

We now consider $a_0 = M^2g^2r \cos \theta_0(-y_0 \sin \phi_0 + x_0 \cos \phi_0)$. Given the values of θ_0, ϕ_0, and y_0, this becomes $a_0 = -M^2g^2rx_0 \cdot a_0$ is definitely negative.

The final result for case 1 is that the first column of the Routh table (see Table 5.5) starts with $a_4 > 0$, sees no sign change as we move down the first column to $a_2 > 0$ and sees one sign change as we move down to $a_0 < 0$. As a result, there is one eigenvalue with a positive real part (one sign change) and, therefore, one unstable mode. This is the same result we obtained from the numerical study. The difference is that applying the Routh stability criterion does not give us information related to "how unstable the mode is" (i.e. where the eigenvalue with the positive real value lies on the complex plane) or on the mode shape that is unstable.

We now consider case 2 where $\theta_0 = 0$ and $\phi_0 = 0$. This time, again with $y_0 = 0$, we find,

$$a_2 = Mgr[I \cos \theta_0 + Mr(-y_0 \sin \phi_0 + x_0 \cos \phi_0)] = Mgr(I + Mrx_0).$$

---

13 The reader is encouraged to review the definitions of moments and products of inertia in Appendix B. The *parallel axis theorem* states that the moment of inertia about some point B on a rigid body is equal to the moment of inertia about the center of mass, I_G, plus the mass of the rigid body, M, multiplied by the square of the distance between G and B. That is, in this case, $I = I_G + Mx_0^2$ or $I_G = I - Mx_0^2$.

In this case, $a_2 > 0$ for all parameter values.

In case 2, $a_0 = M^2 g^2 r \cos \theta_0 (-y_0 \sin \phi_0 + x_0 \cos \phi_0) = M^2 g^2 r x_0$, which is greater than zero for all parameter values. Therefore, case 2 sees positive values for all elements in the first column of the Routh table and the equilibrium position is stable as we predicted numerically.

Consideration of cases 3 and 4 follow directly from what has just been presented.

In case 3, $a_4 > 0$, $a_2 = Mgr[I \cos \theta_0 + Mr(-y_0 \sin \phi_0 + x_0 \cos \phi_0)] = Mgr(-I + Mrx_0) < 0$, and $a_0 = M^2 g^2 r \cos \theta_0 (-y_0 \sin \phi_0 + x_0 \cos \phi_0) = -M^2 g^2 r x_0 < 0$ so that there is one sign change and one unstable mode.

Similarly, in case 4, $a_4 > 0$, $a_2 = Mgr[I \cos \theta_0 + Mr(-y_0 \sin \phi_0 + x_0 \cos \phi_0)] = Mgr(-I - Mrx_0) < 0$, and $a_0 = M^2 g^2 r \cos \theta_0 (-y_0 \sin \phi_0 + x_0 \cos \phi_0) = M^2 g^2 r x_0 > 0$ so that there are two sign changes and two unstable modes.

Notice that case 2 is stable and case 4 has two sign changes, and therefore two unstable modes, for any set of parameter values. Cases 1 and 3 have the option of changing the sign of a_2 depending on whether $(I - Mrx_0) > 0$ or not. In fact, if $(I - Mrx_0)$ changes sign because different parameter values are used, the result is simply a shift of where the sign change occurs in the first column of the Routh array and not a change in the number of sign reversals. As a result, cases 1 and 3 have a single sign change and therefore one unstable mode regardless of the sign of a_2.

5.5 Standard Procedure for Stability Analysis

We start by assuming that the equations of motion for a system have been derived, an equilibrium solution has been found, and the equations of motion have been linearized for small motions about this equilibrium position to give N second order linear differential equations for the N degrees of freedom, as follows.

$$[M]\{\ddot{x}\} + [C]\{\dot{x}\} + [K]\{x\} = \{0\}. \tag{5.29}$$

Whether or not the system is stable for small perturbations away from the equilibrium operating point requires that the eigenvalues and eigenvectors (if we are interested in mode shapes) of these differential equations be found.

You will find that computer programs for finding eigenvalues work with the standard eigenvalue problem, which starts with a set of first-order differential equations,

$$\{\dot{x}\} = [A]\{x\} \tag{5.30}$$

and then assumes exponential solutions $\{x(t)\} = \{X\}e^{\lambda t}$ to arrive at, after exchanging sides in the equation, the standard eigenvalue problem,

$$[A]\{X\} = \lambda\{X\}. \tag{5.31}$$

The problem is then one of transforming Equation 5.29 to a form equivalent to that in Equation 5.30. This can be accomplished using the following steps.

- Define a new set of variables for the displacements

$$\{y_1\} = \{x\} \Rightarrow \{\dot{y}_1\} = \{\dot{x}\}. \tag{5.32}$$

- Define a second set of new variables for the rates of change of the displacements

$$\{y_2\} = \{\dot{x}\} \Rightarrow \{\dot{y}_2\} = \{\ddot{x}\}.$$ (5.33)

- From the preceding two definitions, we can see that we have established two expressions for the rates of change of the displacements. These two expressions must agree, resulting in a set of N first-order differential equations,

$$\{\dot{y}_1\} = \{y_2\}.$$ (5.34)

- We still have our initial differential equations (Equation 5.29)

$$[M]\{\ddot{x}\} + [C]\{\dot{x}\} + [K]\{x\} = \{0\}$$

which, after noting that $\{\ddot{x}\} = \{\dot{y}_2\}$, can be written as,

$$[M]\{\dot{y}_2\} + [C]\{y_2\} + [K]\{y_1\} = \{0\}$$

or

$$\{\dot{y}_2\} = -[M]^{-1}[K]\{y_1\} - [M]^{-1}[C]\{y_2\}.$$ (5.35)

- Equations 5.34 and 5.35 are then written together as a set of $2N$ differential equations.

$$\left\{ \begin{array}{c} \dot{y}_1 \\ \dot{y}_2 \end{array} \right\} = \left[\begin{array}{cc} 0 & I \\ -M^{-1}K & -M^{-1}C \end{array} \right] \left\{ \begin{array}{c} y_1 \\ y_2 \end{array} \right\}.$$ (5.36)

- The $2N \times 2N$ constant coefficient matrix,

$$\left[\begin{array}{cc} 0 & I \\ -M^{-1}K & -M^{-1}C \end{array} \right]$$

is defined to be $[A]$ so that the differential equations can be written as,

$$\{\dot{y}\} = [A]\{y\}$$

which is in the desired form of Equation 5.30.

Standard eigenvalue routines for finding eigenvalues and eigenvectors simply require that the matrix $[A]$ be provided. The user must be careful to determine that the algorithm used is for *non-symmetric matrices* when sending a typical $[A]$ matrix from a dynamic system since in our case $[A]$ is most often non-symmetric. Computer routines for finding the eigenvalues and eigenvectors of symmetric matrices are much more efficient than those for non-symmetric matrices and are very common but they cannot process non-symmetric matrices. It is up to the user to determine which algorithm to use in any given software package.

If used properly, the computer package will return a set of $2N$ complex eigenvalues of the form,

$$\lambda_j = \sigma_j + i\omega_j; \; j = 1, 2N.$$

We have previously noted that the stability of the system is dependent upon the signs of the σ_j and the frequencies of oscillation on the ω_j.

For each eigenvalue there will be a corresponding eigenvector. This means that $2N$ eigenvectors will be returned, each consisting of a column of $2N$ complex values. They contain valuable information about the *mode shapes* inherent in the system and deserve a little extra explanation which will be presented in the next chapter.

Exercises

Descriptions of the systems referred to in the exercises are contained in Appendix A.

5.1   Let an equilibrium position for system 1 be θ_{1_0} and θ_{2_0}. Show that the mass and stiffness matrices for small motions about the equilibrium position are,

$$[M] = \begin{bmatrix} (m_1 + m_2)d_1^2 + m_2 d_2^2 + 2m_2 d_1 d_2 \cos\theta_{2_0} & m_2 d_2^2 + m_2 d_1 d_2 \cos\theta_{2_0} \\ m_2 d_2^2 + m_2 d_1 d_2 \cos\theta_{2_0} & m_2 d_2^2 \end{bmatrix}$$

$$[K] = \begin{bmatrix} -(m_1 + m_2)gd_1 \sin\theta_{1_0} - m_2 gd_2 \sin(\theta_{1_0} + \theta_{2_0}) & -m_2 gd_2 \sin(\theta_{1_0} + \theta_{2_0}) \\ -m_2 gd_2 \sin(\theta_{1_0} + \theta_{2_0}) & -m_2 gd_2 \sin(\theta_{1_0} + \theta_{2_0}) \end{bmatrix}.$$

5.2   Consider system 1 for the case where $m_1 = m_2 = m$ and $d_1 = d_2 = d$. Write the characteristic equation for small motions about each of the four equilibrium states found in Exercise 4.1. Determine the stability of each equilibrium state using the Routh stability criterion and confirm the results by finding the eigenvalues.

5.3   Linearize the equation of motion for system 4 about the equilibrium position determined in Exercise 4.2 and discuss the stability for: (1) $(k - m\,\Omega^2) < 0$; (2) $(k - m\,\Omega^2) = 0$; and (3) $(k - m\,\Omega^2) > 0$.

5.4   Linearize the equation of motion for system 6 around the two equilibrium positions found in Exercise 4.3 and determine whether or not they are stable.

5.5   Two equilibrium states were found for system 8 in Exercise 4.5. Comment on the stability of the two states. If there is an instability, specify the type.

5.6   Are the two equilibrium states found for system 9 in Exercise 4.6 stable?

5.7   Comment on the stability of the four equilibrium states found for system 17 in Exercise 4.8.

6

Mode Shapes

This is a separate chapter on *mode shapes*, included because of the importance of mode shapes in understanding the dynamic behavior of systems. Assume that the analyst has created a dynamic model of a system, found the equilibrium states, linearized the equations of motion about one of them, and tested that state for stability. In the case where the eigenvalues indicate an instability, the question becomes – *what design changes can be made to make the system stable?* The answer may be as simple as adding stiffness or damping somewhere in the system – *but where?* The mode shapes answer this question.

Equation 5.13 in Section 5.1 shows that the total response of the linear system is a combination of all of the natural modes. The job of the analyst is to identify which mode is unstable, visualize the body as it moves in that mode and then choose locations where there are relatively large deflections between bodies as the locations to attempt to stabilize the system by adding stiffness or damping. Adding stiffness or damping between two points that have no relative motion in this mode shape will be fruitless.

Visualization of mode shapes is much enhanced by computer animation but the effort involved in doing that is often not justified. Quick, hand-drawn sketches can be just as effective in achieving the design goals.

6.1 Eigenvectors

The eigenvalues and eigenvectors are calculated for a general dynamic system that has been cast into a first-order form as shown in Equation 5.36 and our interpretation of mode shapes begins with that general system. An eigenvalue $\lambda_j = \sigma_j + i\omega_j$ will be accompanied by an eigenvector,

$$\left\{ \begin{array}{c} Y_1 \\ Y_2 \end{array} \right\}_j$$

which contains, in its first N elements, the complex amplitudes of $\{y_1(t)\}$ and, in its last N elements, the complex amplitudes of $\{y_2(t)\}$. In Equations 5.32 and 5.33 we defined $\{y_1(t)\}$ as the original set of displacement degrees of freedom, $\{x(t)\}$, and $\{y_2(t)\}$ as the derivatives of $\{x(t)\}$. That is, $\{y_1(t)\} = \{x(t)\}$ and $\{y_2(t)\} = \{\dot{x}(t)\}$. We therefore expect the first N elements of the eigenvector to represent the displacements in the system and to be very useful

The Practice of Engineering Dynamics, First Edition. Ronald J. Anderson.
© 2020 John Wiley & Sons Ltd. Published 2020 by John Wiley & Sons Ltd.
Companion Website: www.wiley.com/go/anderson/engineeringdynamics

in describing how the system degrees of freedom move in relation to each other. The last half of the eigenvector contains derivatives of the first half that, given that we assumed exponential solutions of the form $\{x(t)\} = \{X\}e^{\lambda t}$, can be regenerated simply by multiplying the first N elements of the eigenvector by λ_j since $\{\dot{x}(t)\} = \lambda\{X\}e^{\lambda t} = \lambda\{x(t)\}$.

In other words, the eigenvector returned by whatever computer package is being used is actually,

$$\left\{ \begin{array}{c} Y_1 \\ Y_2 \end{array} \right\}_j = \left\{ \begin{array}{c} X \\ \dot{X} \end{array} \right\}_j = \left\{ \begin{array}{c} X \\ \lambda X \end{array} \right\}_j. \tag{6.1}$$

It is convenient therefore to discard the last N elements of the eigenvector and work with the displacements contained in the first N elements.

Consider now an eigenvector of length N, the elements of which are all complex numbers representing the displacement degrees of freedom in a system. Such a vector would look like,

$$\{X\}_j = \left\{ \begin{array}{c} a_1 + ib_1 \\ a_2 + ib_2 \\ \vdots \\ a_p + ib_p \\ \vdots \\ a_n + ib_n \\ \vdots \\ a_N + ib_N \end{array} \right\}_j \tag{6.2}$$

where the coefficients a_p and b_p, for all p, are real valued constants.

Remembering that the elements in the eigenvector are of arbitrary magnitude since they represent only the relationship of one degree of freedom to another, there is no loss of information[1] if we normalize the elements of the eigenvector by dividing by the element of largest absolute value. Assume that the element $a_n + ib_n$ has the largest magnitude and divide all of the others by it. Note that this involves dividing by a complex number, $a_n + ib_n$, and not by the magnitude of the number, $\sqrt{a_n^2 + b_n^2}$. The eigenvector is transformed to,

$$\{X\}_j = \left\{ \begin{array}{c} c_1 + id_1 \\ c_2 + id_2 \\ \vdots \\ c_p + id_p \\ \vdots \\ 1 + i0 \\ \vdots \\ c_N + id_N \end{array} \right\}_j \tag{6.3}$$

---

1 This was shown in Section 5.1 where we noted that we could never solve for all of the elements of $\{X\}$ uniquely since there were only N equations to work with and there were $N + 1$ unknowns – the eigenvalue and the N elements of the eigenvector. This meant that we could find only the eigenvalue and $N - 1$ other pieces of information that are components of the eigenvector relative to each other. As a result, we can scale all of the elements of an eigenvector by the same factor without losing any information.

where c_p and d_p are real valued constants and the nth degree of freedom, which is now represented by $1 + i0$, becomes the reference degree of freedom as we move forward from here.

We now transform the eigenvector one more time by changing each element to an amplitude, $|X_p| = \sqrt{c_p^2 + d_p^2}$, and a phase angle with respect to the degree of freedom with maximum amplitude, $\phi_p = \arctan(d_p/c_p)$. The final form of the eigenvector is then,

$$\{X\}_j = \begin{Bmatrix} |X_1| \angle \phi_1 \\ |X_2| \angle \phi_2 \\ \vdots \\ |X_p| \angle \phi_p \\ \vdots \\ 1 \angle 0 \\ \vdots \\ |X_N| \angle \phi_N \end{Bmatrix}_j \tag{6.4}$$

where the expression $\angle \phi_p$ means "at an angle ϕ_p" with respect to the phase angle of the reference degree of freedom.

The reason for producing the eigenvectors is so that we can observe the natural movements of the system as functions of time. To this point, we have generated only an amplitude and a phase angle with respect to the degree of freedom having the largest amplitude. Since we cannot have an eigenvector without having first found the corresponding eigenvalue, we also have some idea of the time dependence of the modes of motion. That is, we know if a mode oscillates or not and we know how much damping each mode has. In general, we are mostly interested in the mode shapes of those modes that are unstable or nearly unstable. Any corrective action we can take to improve the stability of these modes depends on visualizing the motions and applying force-producing elements to areas in the system where they will have some effect. For example, if an unstable mode involves motions where two adjacent bodies have no relative motion with respect to each other, there is little to be gained from placing a spring or damper between these two bodies. Such an element will produce no force and, as a result, make no contribution to stabilizing the system.

We know that the motion resulting from mode-j of the system can be written, as a function of time, as $\{x(t)\} = \{X\}_j e^{\lambda_j t}$ or, given that $\lambda_j = \sigma_j + i\omega_j$ and substituting $e^{\lambda_j t} = e^{\sigma_j t}(\cos \omega_j t + i \sin \omega_j t)$, we can write $\{x(t)\} = \{X\}_j e^{\sigma_j t}(\cos \omega_j t + i \sin \omega_j t)$.

In order to visualize the motion, we normally ignore the damping term, $e^{\sigma_j t}$, for the simple reason that it causes the amplitudes to grow or decay with time and therefore clouds our image of the motion. We already know if the mode is unstable or not, we simply want to see what type of motion is involved.

The apparent conflict between the "real" and "imaginary" parts of the response has been partly resolved by transforming the eigenvector into an amplitude and a phase angle. Given that we are ignoring the decay or growth for the purpose of visualization, we can write the time variation of the system in the *j*th mode as,

$$\{x(t)\} = \begin{Bmatrix} |X_1|(\cos(\omega_j t + \phi_1) + i\sin(\omega_j t + \phi_1)) \\ |X_2|(\cos(\omega_j t + \phi_2) + i\sin(\omega_j t + \phi_2)) \\ \vdots \\ |X_p|(\cos(\omega_j t + \phi_p) + i\sin(\omega_j t + \phi_p)) \\ \vdots \\ 1(\cos(\omega_j t) + i\sin(\omega_j t)) \\ \vdots \\ |X_N|(\cos(\omega_j t + \phi_N) + i\sin(\omega_j t + \phi_N)) \end{Bmatrix}. \tag{6.5}$$

The real and imaginary parts of Equation 6.5 differ only in that each is the projection of the complex function on a different axis of the complex plane. We can pick either the real or the imaginary component to describe the motion since each has the same amplitude and frequency. Without loss of generality, we can choose to watch the motion as it is projected on the imaginary axis and write,

$$\{x(t)\} = \begin{Bmatrix} |X_1|\sin(\omega_j t + \phi_1) \\ |X_2|\sin(\omega_j t + \phi_2) \\ \vdots \\ |X_p|\sin(\omega_j t + \phi_p) \\ \vdots \\ 1\sin(\omega_j t) \\ \vdots \\ |X_N|\sin(\omega_j t + \phi_N) \end{Bmatrix}. \tag{6.6}$$

Table 5.2 lists the eigenvalues and eigenvectors for the two degrees of freedom system we have considered so many times. Here, we are interested in case 2 – the equilibrium state with two stable eigenvalues. From the table, there is a mode with $\lambda = \pm 4.6970i$. This mode[2] has a frequency of 4.6970 rad s^{-1} or 0.746 Hz and is described by a motion where the angle[3] ϕ is roughly 1/4 (actually 0.2481) of the magnitude of the angle θ and is out-of-phase (the negative sign on the second element of the eigenvector indicates a 180° phase angle) with θ. Figure 6.1 shows the system in equal increments of time as it moves in this mode. The

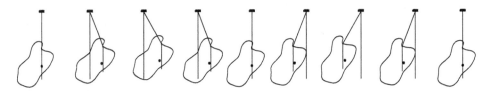

Figure 6.1 The higher frequency mode for the two degrees of freedom system (0.746 Hz).

---

2 Table 5.2 actually shows two modes with similar eigenvalues, one with $\lambda = +4.6970i$ and another with $\lambda = -4.6970i$. Since eigenvalues with non-zero imaginary parts appear as complex conjugate pairs, these are actually the same mode.
3 The degrees of freedom, θ and ϕ are defined in Figure 3.6.

Figure 6.2 The lower frequency mode for the two degrees of freedom system (0.314 Hz).

motion consists mainly of rotation of the supporting rod with the rigid body maintaining nearly the same orientation at all times as it swings back and forth.

Figure 6.2 shows the motion for the second mode associated with the equilibrium of case 2. From Table 5.2, we see that this is a lower frequency mode than the first (1.9746 rad s^{-1} or 0.314 Hz) and is dominated by the angle ϕ. θ has roughly 10% (actually 0.1110) of the amplitude of ϕ and the angles are in-phase since both elements have the same sign. We expect a motion where the supporting rod moves very little and the rigid body essentially swings about the supporting point B. This motion is shown in Figure 6.2 and confirms the point just made.

6.2 Comparing Translational and Rotational Degrees of Freedom

The example just presented did not contain an important element in the interpretation of mode shapes. In looking at the example we were comparing two rotational degrees of freedom, θ and ϕ, and were able to make, and understand, statements like "θ has roughly 10% (actually 0.1110) of the amplitude of ϕ" and understand what they mean. The equations of motion for many systems have both rotational and translational degrees of freedom. What does it mean to say that "x is 10% of α" when x is a displacement and α is an angle?

To answer this question, we use another example. Figure 6.3 shows a thin, uniform, rigid rod of length $2a$ that is suspended at its ends by two springs of stiffness k_1 and k_2. The first view in the figure shows the system in its equilibrium position. The rod has mass m and a moment of inertia about its center of mass of $ma^2/3$. A linear damper with coefficient c is included at a distance b from the center of mass as shown.

We let the rod have two degrees of freedom. Its center of mass is free to move up and down and we designate this motion by the variable x, which is positive upwards and is a measure of how far the center of mass moves away from its equilibrium position. In addition, the rod can rotate in the plane of its motion and we use the angle α to designate the rotation away from the equilibrium orientation.

The kinematic analysis of this system is very simple. The acceleration of the center of mass is $\ddot{x}$ vertically upward and the rod has an angular acceleration of $\ddot{\alpha}$ in the direction shown for positive α in Figure 6.3.

Figure 6.3 also shows the free-body diagram for the system. Notice that the weight of the rod is missing since it will be offset by preloads in the springs[4]. As a result, we also don't

---

4 Finding eigenvalues and eigenvectors requires that the linear equations of motion about an equilibrium state be derived. If this is all that is required, there is no need to derive the fully nonlinear equations of

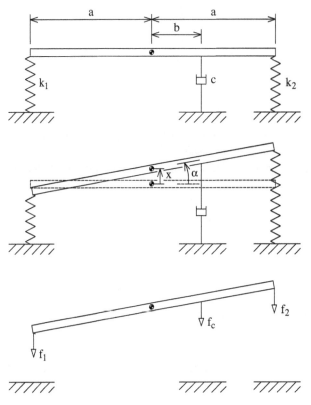

Figure 6.3 A rigid rod supported on springs.

include the preloads when we establish expressions for the spring forces f_1 and f_2. They are only the changes in the spring forces that result from displacements away from their equilibrium displacements.

Using the symbol δ to designate the displacements of the elements from their equilibrium lengths, we can write,

$$\delta_1 = x - a\alpha$$
$$\delta_2 = x + a\alpha$$
$$\delta_c = \dot{x} + b\dot{\alpha} \tag{6.7}$$

---

motion, establish the equilibrium positions, and then linearize the equations of motion about that equilibrium. We simply assume that the system is in an easily determined equilibrium (e.g. resting on the springs in this case) and then derive the linear equations of motion for motions away from equilibrium directly. This is the method most often used in the study of *vibrations* as opposed to *dynamics*. Vibrations is primarily concerned with the linear, frequency domain, response of systems. One of the subtleties in deriving these linear equations of motion is that the forces that acted to put the system in the equilibrium position are canceled out by constant forces in elements such as springs and both are ignored in determining the equations of motion. In the case considered here, gravity acted on the rod to deflect the springs downward until an equilibrium was established. As we consider motions away from that equilibrium, we consider neither gravity nor the preloads in the springs.

where the assumption of small motions has been used implicitly. That is, no trigonometric functions of the angle α appear because the small angle assumptions $\sin \alpha \approx \alpha$ and $\cos \alpha \approx 1$ have been used.

A short discussion of the directions assumed for the forces in the elements is in order at this point because it is very easy to lose a sign. Construction of the free-body diagram in Figure 6.3 was not possible without deciding on directions for the forces acting in the springs and the damper. The spring forces, f_1 and f_2, are shown as if the springs are in tension. That is, the assumption is that the motions of the system have caused both springs to become longer than they were in the equilibrium position and the springs are both trying to pull the rod down in order to return to equilibrium. Whether or not this is actually the case depends on what the actual motion of the system is and we cannot predict that without first deriving the equations of motion. Assuming tension in both springs is as good an assumption as any so long as we maintain that assumption for all the following steps in the derivation. The drawing in the center of Figure 6.3 seems to imply that if the spring on the right hand end is in tension than the spring on the left must be in compression. Again, we have no way of saying in advance whether this is so or not. The center drawing is simply there to show the positive directions assumed for the degrees of freedom, x and α. No prior knowledge of the motion was available when it was drawn. The point is that the degrees of freedom have been assigned positive directions and assumptions were made about the displacements of the springs. The analyst must respect these choices when deriving the equations of motion.

With respect to the force in the damper, f_c, the same arguments as were made for the springs apply. Linear dampers resist getting longer or shorter by virtue of a constitutive relationship that says that $f_c = c\dot{\delta}$ where c is a constant damping coefficient. The free-body diagram shows that f_c has been assumed to be acting downward on the rod. This can only happen if the damper is getting longer so the expression for $\dot{\delta}_c$ in Equation 6.7 has been derived so that it is positive if the damper is increasing in length.

A final note about Equation 6.7 is that the expressions presented respect the assumed directions of the forces and use the positive senses of the degrees of freedom to calculate the corresponding spring displacements and damper velocity.

The forces acting can then be written as,

$$f_1 = k_1 \delta_1 = k_1(x - a\alpha)$$
$$f_2 = k_2 \delta_2 = k_2(x + a\alpha)$$
$$f_c = c\dot{\delta}_c = c(\dot{x} + b\dot{\alpha}). \tag{6.8}$$

We now sum forces vertically upward. Remember that x was assumed to be positive upward and that positive acceleration is in the same direction as x. According to Newton's second law, there can only be an acceleration in that direction if there is a net force in that direction. The implication is that some or all of the force directions we assumed were incorrect but consistency in applying our assumptions will lead to the correct equations of motion. The force balance results in,

$$m\ddot{x} = -f_1 - f_2 - f_c$$
$$= -k_1(x - a\alpha) - k_2(x + a\alpha) - c(\dot{x} + b\dot{\alpha})$$
$$= -c\dot{x} - cb\dot{\alpha} - (k_1 + k_2)x + a(k_1 - k_2)\alpha. \tag{6.9}$$

Summing moments about the center of mass in the direction of positive $\ddot{\alpha}$ (i.e. the same counter-clockwise direction as positive α) yields,

$$\frac{ma^2}{3}\ddot{\alpha} = af_1 - af_2 - bf_c$$

$$= ak_1(x - a\alpha) - ak_2(x + a\alpha) - bc(\dot{x} + b\dot{\alpha})$$

$$= -bc\dot{x} - cb^2\dot{\alpha} + a(k_1 - k_2)x - a^2(k_1 + k_2)\alpha. \tag{6.10}$$

Combining Equations 6.9 and 6.10 into the standard matrix form gives,

$$\begin{bmatrix} m & 0 \\ 0 & \frac{1}{3}ma^2 \end{bmatrix}\begin{Bmatrix} \ddot{x} \\ \ddot{\alpha} \end{Bmatrix} + \begin{bmatrix} c & cb \\ cb & cb^2 \end{bmatrix}\begin{Bmatrix} \dot{x} \\ \dot{\alpha} \end{Bmatrix}$$

$$+ \begin{bmatrix} k_1 + k_2 & -a(k_1 - k_2) \\ -a(k_1 - k_2) & a^2(k_1 + k_2) \end{bmatrix}\begin{Bmatrix} x \\ \alpha \end{Bmatrix} = \begin{Bmatrix} 0 \\ 0 \end{Bmatrix}. \tag{6.11}$$

We consider two cases. The first will be a case with no damping (i.e. $c = 0$) in order to demonstrate a mode shape where we compare a translation (x) with a rotation (α). The second will include a significant amount of damping so that the phase angles between degrees of freedom becomes obvious.

Let the parameter values be as shown in Table 6.1.

The eigenvalues and corresponding eigenvectors for the first case are,

$$\lambda_1 = \pm 11.652i \; ; \{X\}_1 = \begin{Bmatrix} 1.0000 \\ -0.1056 \end{Bmatrix}$$

and

$$\lambda_2 = \pm 20.597i \; ; \{X\}_2 = \begin{Bmatrix} 0.1407 \\ 1.0000 \end{Bmatrix}.$$

The first mode is therefore an undamped oscillation at a frequency of 1.85 Hz (11.652 rad s^{-1}) with the amplitude of x dominating. α is roughly 10% of x (actually 10.56%) and has the opposite sign. The second mode is another undamped oscillation (no surprise since $c = 0$ for this case), this time at 3.28 Hz (20.597 rad s^{-1}) with x being only 14% of α. The question is, what do these mode shapes look like? How do we compare translation to rotation?

Table 6.1 Parameters for the rigid rod example.

Parameter	Case 1	Case 2
m	1000 kg	1000 kg
a	2 m	2 m
b	1 m	1 m
k_1	60000 Nm^{-1}	60000 Nm^{-1}
k_2	80000 Nm^{-1}	80000 Nm^{-1}
c	0 N ms^{-1}	5000 N ms^{-1}

Visualization of the mode shapes is akin to drawing the system as it moves in time. To draw the rigid rod, we first need to know where the center of mass is (i.e we need to know x). After making this translation, we need to know where the ends of the rod are located and we do this using the same linear equations we used to find spring deflections (Equation 6.8). It is clear that the positions of the ends of the rod relative to the position of the center of mass depend on the angle α and the distance from the center of mass to the ends of the rod. The absolute vertical positions of the rod ends (written as $p_{\text{left}}^{\text{v}}$ and $p_{\text{right}}^{\text{v}}$) can be expressed as,

$$p_{\text{left}}^{\text{v}} = x - a\alpha$$
$$p_{\text{right}}^{\text{v}} = x + a\alpha. \tag{6.12}$$

Given the positions of the ends of the rod as a function of time or, more precisely, as a function of the what proportion of a cycle of motion has passed, we can connect the ends with a straight line and visualize the motion. The expressions for the motions of the ends of the rod (substituting the eigenvectors for the two mode shapes and $a = 2$ m from Table 6.1) are, for the first mode,

$$p_{\text{left}}^{\text{v}} = [1.000 - 2.0(-0.1056)] \sin(11.652t)$$
$$= 1.2112 \sin(11.652t)$$
$$p_{\text{right}}^{\text{v}} = [1.000 + 2.0(-0.1056)] \sin(11.652t)$$
$$= 0.7888 \sin(11.652t) \tag{6.13}$$

and, for the second,

$$p_{\text{left}}^{\text{v}} = [0.1407 - 2.0(1.0000)] \sin(20.597t)$$
$$= -1.8593 \sin(20.597t)$$
$$p_{\text{right}}^{\text{v}} = [0.1407 + 2.0(1.0000)] \sin(20.597t)$$
$$= 2.1407 \sin(20.597t). \tag{6.14}$$

Having generated the expressions for the motions, presented as Equations 6.13 and 6.14, one can draw the rod for various values of ωt in the range $0 < \omega t < 2\pi$, thereby showing a complete cycle of the harmonic motion. Figure 6.4 shows the two modes. Remember that the amplitudes given in Equations 6.13 and 6.14 are not absolute since the expressions are derived from the eigenvectors that contain only relative amplitude information. A sketch of a mode shape can use any convenient value for the vertical displacement of one end of the rod. The motion of the other end must satisfy the relative amplitudes given in the equations.

6.3 Nodal Points in Mode Shapes

In Section 6.2, the vertical motions of the left and right ends of the rigid rod were given in Equation 6.12. If you compare these to the expressions for the spring deflections at the ends of the rod (see Equation 6.7), you will see that the expressions are identical. Basically, the deflection of a spring connected from the moving body to the ground is a measure of the motion of the point on the body relative to ground (i.e. absolute motion) in the direction

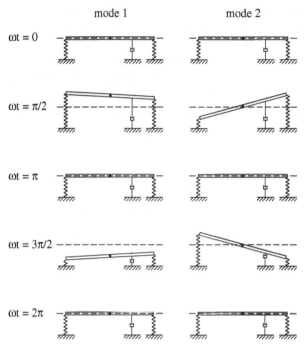

Figure 6.4 Undamped mode shapes for the rigid rod supported on springs.

aligned with the spring. We can easily write an expression for the deflection of a spring connecting the rod to the ground at some distance, ℓ, to the right of the center of mass as,

$$\delta = x + \ell\alpha \tag{6.15}$$

and, if we are looking for a point that has no motion at all, we simply set the deflection of the spring to zero and solve for the location of that point. For $\delta = 0$, we get

$$\ell = \frac{-x}{\alpha} \tag{6.16}$$

which says that the point with no motion in the first mode is located at $\ell = 9.4697$ m and, in the second mode, at $\ell = -0.1407$ m.

Figure 6.5 shows the nodal points for the rigid rod. Nodal points are useful aids in visualizing mode shapes. The motion of the system is simply a rotation about the nodal points.

6.4 Mode Shapes with Damping

In Table 6.1, we introduced a second case for the rigid rod supported on springs that we have been considering. The second case included significant damping and the calculated eigenvalues and eigenvectors for that case are,

$$\lambda_1 = -2.1045 \pm 11.889i; \quad \{X\}_1 = \left\{ \begin{array}{c} -0.0144357 + 0.0815538i \\ -0.0122252 - 0.0086485i \end{array} \right\}$$

Figure 6.5 Nodal points.

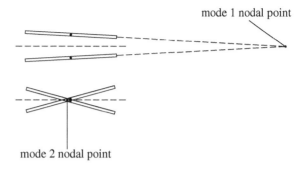

mode 1 nodal point

mode 2 nodal point

and,

$$\lambda_2 = -2.2705 \pm 19.747i; \; \{X\}_2 = \left\{ \begin{array}{c} 0.0188754 + 0.0071370i \\ -0.0057466 + 0.0499797i \end{array} \right\}.$$

These are in the raw form of Equation 6.2.

The element having the largest amplitude in each eigenvector is selected and divided into the other element(s) using complex division. In the first case, the first element of $\{X\}_1$ is the largest so we divide by $-0.0144357 + 0.0815538i$ to get,

$$\lambda_1 = -2.1045 \pm 11.889i; \; \{X\}_1 = \left\{ \begin{array}{c} 1.0000000 + 0.0000000i \\ -0.0770966 + 0.1635498i \end{array} \right\}.$$

Similarly, we divide $\{X\}_2$ throughout by the first element, (i.e. $0.0188754 + 0.0071370i$, to get

$$\lambda_2 = -2.2705 \pm 19.747i; \; \{X\}_2 = \left\{ \begin{array}{c} 0.0980779 - 0.3889390i \\ 1.0000000 + 0.0000000i \end{array} \right\}.$$

These are now in the form of Equation 6.3.

The final step is to change the elements of the eigenvectors to magnitudes and phase angles as in Equation 6.4. This results in,

$$\lambda_1 = -2.1045 \pm 11.889i; \; \{X\}_1 = \left\{ \begin{array}{c} 1.0000 \angle 0.00° \\ 0.1808 \angle -115.24° \end{array} \right\}.$$

and,

$$\lambda_2 = -2.2705 \pm 19.747i; \; \{X\}_2 = \left\{ \begin{array}{c} 0.4011 \angle 75.85° \\ 1.0000 \angle 0.00° \end{array} \right\}.$$

Using Equation 6.6, we can write the time variation of the mode shapes as,

$$\{x(t)\}_1 = \left\{ \begin{array}{c} 1.0000 \sin(\omega_1 t) \\ 0.1808 \sin(\omega_1 t - 115.24°) \end{array} \right\}$$

and,

$$\{x(t)\}_2 = \left\{ \begin{array}{c} 0.4011 \sin(\omega_2 t + 75.85°) \\ 1.0000 \sin(\omega_2 t) \end{array} \right\}$$

where, from the eigenvalues, $\omega_1 = 11.889$ rad s^{-1} (1.89 Hz) and $\omega_2 = 19.747$ rad s^{-1} (3.14 Hz).

In fact, the actual values of the natural frequencies, while useful, are not important to visualizing the mode shapes. We simply want to plot the system's position for fractions of a period of motion, $0 \le \omega t \le 2\pi$, so we use a set of values of $\omega_1 t$ and $\omega_2 t$ between 0 and 2π. We are basically creating frames of an animation when we do this. If you are actually doing computer animation, you use as many frames as necessary to get a smooth animation. When using paper, as we are here, it is generally sufficient to use the set, $\omega t = 0, \pi/2, \pi, 3\pi/2, 2\pi$.

To make the plotting easier, we can use the sine of the sum of angles identity to write

$$\sin(\omega t + \phi) = \sin \omega t \cos \phi + \cos \omega t \sin \phi$$

so that the time variation of the mode shapes becomes

$$\{x(t)\}_1 = \left\{ \begin{array}{c} 1.0000 \sin \omega_1 t \\ 0.1808 \sin \omega_1 t \cos(-115.24°) + 0.1808 \cos \omega_1 t \sin(-115.24°) \end{array} \right\}$$

and,

$$\{x(t)\}_2 = \left\{ \begin{array}{c} 0.4011 \sin \omega_2 t \cos 75.85° + 0.4011 \cos \omega_2 t \sin 75.85° \\ 1.0000 \sin \omega_2 t \end{array} \right\}.$$

Substituting numerical values for the sines and cosines of the phase angles gives

$$\{x(t)\}_1 = \left\{ \begin{array}{c} 1.0000 \sin \omega_1 t \\ -0.0771 \sin \omega_1 t - 0.1635 \cos \omega_1 t \end{array} \right\}$$

and,

$$\{x(t)\}_2 = \left\{ \begin{array}{c} 0.0981 \sin \omega_2 t + 0.3889 \cos \omega_2 t \\ 1.0000 \sin \omega_2 t \end{array} \right\}$$

which are readily evaluated for the values of ωt we have chosen to use.

Figure 6.6 shows the damped mode shapes. The concept of nodal points (Section 6.3) does not exist for heavily damped systems. Using these two damped mode shapes in Equation 6.16 in order to find the nodal point leaves you with tangent or cotangent functions of ωt that go off to infinity more than once in the range $0 \le \omega t \le 2\pi$ so that the nodal points are undefined.

6.5 Modal Damping

The last concept to discuss with respect to the natural modes of a system is how damping can be quantified in an understandable way. Saying that the first mode in the example just considered has a real part of -2.2705 doesn't give much information about how long we would expect the mode to sustain motion once it starts. That is, if we disturb the system so

Figure 6.6 Damped mode shapes for the rigid rod supported on springs.

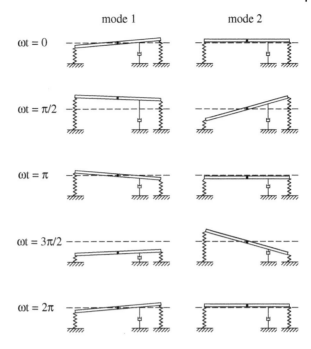

that the response is entirely composed of the first mode, does the amplitude decay[5] within three oscillations or does it take one hundred oscillations?

We start with the equation of motion for a single degree of freedom system for small perturbations about an equilibrium state, which can be written as,

$$m\ddot{x} + c\dot{x} + kx = 0. \tag{6.17}$$

Assuming the usual exponential solution to Equation 6.17,

$$x(t) = Xe^{\lambda t} \tag{6.18}$$

this can be rewritten as,

$$(m\lambda^2 + c\lambda + k)Xe^{\lambda t} = 0. \tag{6.19}$$

The characteristic equation is therefore,

$$m\lambda^2 + c\lambda + k = 0 \tag{6.20}$$

which, after division throughout by m, becomes,

$$\lambda^2 + \frac{c}{m}\lambda + \frac{k}{m} = 0. \tag{6.21}$$

---

5 The exponential decay of oscillations predicted from linear models means that the amplitude never goes to zero but only tends toward zero as time approaches infinity. In fact, there is always some Coulomb friction in systems and the oscillations eventually stop completely because of the work done by friction forces. *Decay* in this section uses the somewhat arbitrary measure of the number of oscillations that reduces the initial amplitude by 90%.

The eigenvalues (i.e. characteristic roots of Equation 6.21) are,

$$\lambda = \frac{1}{2}\left[-\frac{c}{m} \pm \sqrt{\left(\frac{c}{m}\right)^2 - 4\left(\frac{k}{m}\right)}\right].$$

(6.22)

Whether or not the motion is oscillatory at all depends on the sign of the term within the square root of Equation 6.22. This term can be negative, zero, or positive.

If it is negative, the square root will generate an imaginary number and the system will oscillate at the frequency indicated by the result (see the discussion following Equations 5.12 and 5.13). In this case, the system is said to be *underdamped*.

If the term inside the square root is zero, the solution will no longer oscillate. We call the system *critically damped* in this case and define a *critical damping coefficient*, c_{cr}, as the amount of damping which just makes the system stop oscillating. The value,

$$c = c_{cr} = \sqrt{4km}$$

makes the term inside the square root equal to zero.

If the term inside the square root is greater than zero, the system will have a purely exponential behavior and will decay exponentially with time. If the value of c is increased beyond the critical damping value, the system is said to be *overdamped* and the decay will become less rapid with larger values of c.

Consider now introducing two new variables into Equation 6.21. The first is the *undamped natural frequency of the system*[6], $\omega_n = \sqrt{k/m}$. The second is the *damping ratio*, $\zeta = c/c_{cr}$, which is defined to be the ratio of the actual damping in the system to the critical damping.

Making these substitutions into Equation 6.21 leads to the following sequence of equations,

$$\lambda^2 + \frac{c}{m}\lambda + \frac{k}{m} = 0$$

$$\lambda^2 + \frac{c}{c_{cr}}\frac{c_{cr}}{m}\lambda + \omega_n^2 = 0$$

$$\lambda^2 + \zeta\frac{\sqrt{4km}}{m}\lambda + \omega_n^2 = 0$$

$$\lambda^2 + 2\zeta\sqrt{\frac{k}{m}}\lambda + \omega_n^2 = 0$$

$$\lambda^2 + 2\zeta\omega_n\lambda + \omega_n^2 = 0.$$

(6.23)

The eigenvalues of Equation 6.23, plotted in Figure 6.7, are,

$$\lambda = -\zeta\omega_n \pm \omega_n\sqrt{\zeta^2 - 1}.$$

(6.24)

---

6 The undamped natural frequency is found by determining the eigenvalues of the one degree of freedom system without damping,

$$m\ddot{x} + kx = 0$$

The eigenvalues are the complex conjugate pair,

$$\lambda = \pm\sqrt{k/m}\,i$$

which give undamped oscillatory motion at a frequency $\omega_n = \sqrt{k/m}$.

Figure 6.7 Calculating the damping ratio from the location of an eigenvalue.

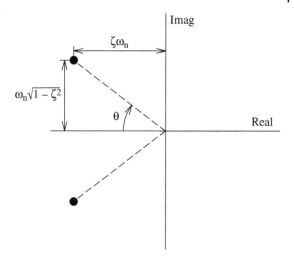

Oscillations will occur when the eigenvalues of Equation 6.24 are a complex conjugate pair. This requires that $\zeta^2 < 1$. That is, the actual damping should be less than critical. If this is the case, then the real part of the eigenvalue is $-\zeta\omega_n$ and the imaginary part is $\pm\omega_n\sqrt{1-\zeta^2}$. Notice that the term $(\zeta^2 - 1)$ has been rewritten as $(1 - \zeta^2)$ because it is now the magnitude of the imaginary part of the eigenvalue and must therefore be positive so that the square root yields a real value. That is, $i = \sqrt{-1}$ has been factored out. The magnitude of the eigenvalue (i.e. the distance from the eigenvalue to the origin in Figure 6.7) is,

$$|\lambda| = \sqrt{\zeta^2\omega_n^2 + \omega_n^2(1 - \zeta^2)} = \omega_n \tag{6.25}$$

and the damping ratio (refer to Figure 6.7) can be found from the trigonometric expression,

$$\cos\theta = \frac{\zeta\omega_n}{\omega_n} = \zeta. \tag{6.26}$$

By *percent damping* of a mode, we mean the percentage of critical damping that is present. Applying the trigonometric analysis just presented to systems with more than one degree of freedom means that any oscillatory eigenvalue can be described by its *percent damping*. For example, for the rigid rod on springs just described, the first mode has 11.4% damping and the second has 17.4% damping[7].

The question now becomes – how many cycles of motion does it take for modes with 11.4% and 17.4% damping to disappear?

---

7 These values are calculated as follows. For the first mode,

$$\zeta = \cos\theta = \frac{2.2705}{\sqrt{(2.2705)^2 + (19.747)^2}} = 0.11423 \approx 11.4\%$$

and, for the second mode,

$$\zeta = \cos\theta = \frac{2.1045}{\sqrt{(2.1045)^2 + (11.889)^2}} = 0.17430 \approx 17.4\%.$$

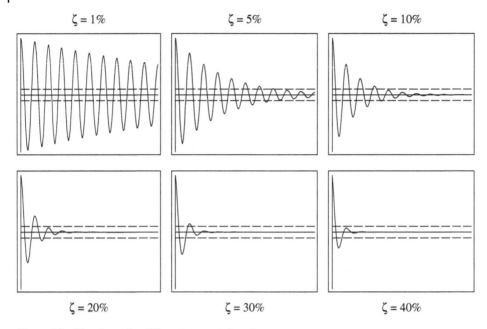

$\zeta = 1\%$ $\zeta = 5\%$ $\zeta = 10\%$

$\zeta = 20\%$ $\zeta = 30\%$ $\zeta = 40\%$

Figure 6.8 Waveforms for different percent damping.

Figure 6.8 shows typical oscillatory behavior of a single degree of freedom system for various levels of percent damping. Also shown on the figure are two horizontal lines representing $\pm 10\%$ of the initial amplitude. It is clear from the figure that oscillations in systems with 30% to 40% damping disappear very rapidly, losing 90% of their amplitude in 1 to 1.5 cycles. With 20% critical damping the oscillation is essentially gone in 2 cycles. A system with 10% critical damping will sustain oscillations for approximately 4 cycles using this measure and those with 1% or 5% will go on for a very long time.

Designers of vehicle suspension systems typically look for modal damping in the 10% to 20% range for the primary modes of a vehicle. Lower levels of damping than this mean that oscillations will be sustained over too long a period for good ride quality. Higher levels of damping will lead to larger shock loads in the system as rapidly changing forces (e.g. from hitting a bump on the road) are transmitted directly to the vehicle by the dampers rather than being attenuated by the springs. Even garage mechanics will check the state of shock absorbers on a car by pushing down a corner of the vehicle and then counting the number of oscillations before the motion stops. If it oscillates more than three times, the shock absorbers need to be replaced.

Exercises

Descriptions of the systems referred to in the exercises are contained in Appendix A.

6.1   Find the two mode shapes for system 10 given that the applied force, $F(t)$, is zero.

6.2 Consider system 13 for the special case where $k=0$ and $d=\ell$. Show that the linearized equations of motion about the equilibrium position, $\theta_0 = 180°$ and $\phi_0 = 0°$, can be written as

$$
\begin{bmatrix} \dfrac{17}{3}m\ell^2 & \dfrac{1}{3}m\ell^2 \\[2mm] \dfrac{1}{3}m\ell^2 & \dfrac{1}{3}m\ell^2 \end{bmatrix} \begin{Bmatrix} \ddot{\alpha} \\ \ddot{\beta} \end{Bmatrix} + \begin{bmatrix} 3mg\ell & 0 \\ 0 & 0 \end{bmatrix} \begin{Bmatrix} \alpha \\ \beta \end{Bmatrix} = \begin{Bmatrix} 0 \\ 0 \end{Bmatrix}
$$

where α and β represent small variations in θ and ϕ respectively.
Find the eigenvalues and eigenvectors of this system and describe in words the two natural modes.

6.3 Derive the equations of motion for system 15 using the horizontal and vertical positions of the mass with respect to point A as degrees of freedom. Find the equilibrium position, linearize about it, and use the mode shapes to show that small horizontal and vertical motions of the mass about equilibrium are independent of each other.

6.4 Find the three mode shapes for system 18 and sketch them.

6.5 A transit system operator is complaining that streetcars (see system 19) manufactured by your company are excessively noisy. The operator has made measurements and cites a peak noise level at 68 Hz as being the main problem. You are told that the cause of the problem is a resonance in the suspension system of the streetcars. Your manager feels that a doubling of the vertical secondary suspension stiffness (k_s) will shift the offending frequency enough that it will no longer be a problem. Your task is to provide analytical evidence to support this solution. Perform the following analysis to do this.

(a) Derive the linearized differential equations of motion governing bounce (x) and pitch (θ) of the truck frame. Use m for the truck frame mass, I for the truck frame pitch moment of inertia, k_s for the secondary stiffness where the truck frame is connected to the car body, and k_p for the primary stiffnesses that connect the wheels to the truck frame. The car body is massive enough that it can be assumed to have no motion and the wheels remain stationary on the rails. Assume that damping is small enough to be neglected. The stable equilibrium state has both bounce and pitch equal to zero.
The result should be,

$$
\begin{bmatrix} m & 0 \\ 0 & I \end{bmatrix} \begin{Bmatrix} \ddot{x} \\ \ddot{\theta} \end{Bmatrix} + \begin{bmatrix} 2k_p + k_s & k_p(b-a) - k_s c \\ k_p(b-a) - k_s c & k_p(a^2+b^2) + k_s c^2 \end{bmatrix} \begin{Bmatrix} x \\ \theta \end{Bmatrix} = \begin{Bmatrix} 0 \\ 0 \end{Bmatrix}.
$$

(b) Find the natural frequencies and mode shapes of the system using, as parameters: $m = 500$ kg, $I = 125$ kg m^2, $a = 1.6$ m, $b = 1.4$ m, $c = 0.1$ m, $k_p = 5 \times 10^6$ N m^{-1}, and $k_s = 1 \times 10^7$ N m^{-1}.

(c) Which mode is causing the problem? Comment on the effectiveness of curing the problem by doubling the secondary suspension stiffness as your manager suggested rather than changing the primary stiffness values. Do not recalculate the natural frequencies. Simply base your comments on the mode shapes.

(d) Add primary and secondary dampers in parallel with the primary and secondary stiffnesses and select damping coefficients so that the two modes each have approximately 15% damping.

6.6 Find the two mode shapes for system 20 and sketch them.

6.7 Find the stable equilibrium position for system 23, linearize the equations of motion around the equilibrium, and find the three natural frequencies and mode shapes.

6.8 In Exercise 4.9, four equilibrium positions were found for the rod in system 12. Derive an expression for the natural frequency of the rod for small motions around the most physically realistic of the equilibrium positions.

6.9 A system with three degrees of freedom has an eigenvalue and corresponding eigenvector as follows:

$$\lambda_1 = -5.1405 + 107.78\ i\ ;\ \{X\}_1 = \begin{Bmatrix} -4.4153 \times 10^{-4} - 9.2574 \times 10^{-3}\ i \\ -9.1116 \times 10^{-4} + 3.0405 \times 10^{-3}\ i \\ +7.0807 \times 10^{-4} - 2.6808 \times 10^{-4}\ i \end{Bmatrix}.$$

Interpret this so as to give the damped natural frequency in Hz, the damping ratio in percent, and a description of the mode shape including amplitudes and phase angles.

7

Frequency Domain Analysis

Determining the response of a mechanical system to a disturbance that can be described primarily by its frequency is often useful. For example, a sensitive piece of equipment may be required to be installed in a power generation plant where the floor is subject to vibrations at the frequency of the rotating machinery that generates the electricity. We must design the new equipment for an environment where it will be disturbed by vibrations having this known frequency but, often, an unknown amplitude. Another use of frequency domain analysis is the case where we have measured the input to a system and have found it to be apparently random. There are well known techniques for decomposing the input into a series of harmonic functions at different frequencies. The task of the analyst is then to predict how the system will respond to the various harmonic excitations and to construct an overall predicted response.

7.1 Modeling Frequency Response

Frequency response calculations use the equations of motion in the linearized form for small motions around a stable equilibrium state. We used the equilibrium states for stability and mode shape analysis in Chapter 5. The difference here is that stability considers the natural motions of the system with no externally applied forces whereas frequency response analysis considers externally applied harmonic forces[1].

Consider the standard set of linearized equations of motion that were presented as Equation 5.29. They are,

$$[M]\{\ddot{x}\} + [C]\{\dot{x}\} + [K]\{x\} = \{0\} \tag{7.1}$$

and they apply to motions of the system away from the equilibrium state *without any applied forces other than those which put the system into the equilibrium state*. We modify the equations by saying that the system is in the equilibrium state but is disturbed by a set of harmonically varying external forces which can be written as,

$$\{F(t)\} = \{F\}e^{i\,\omega t} = \{F\}(\cos \omega t + i \sin \omega t) \tag{7.2}$$

---

[1] As we will see in Chapter 8, only time domain simulations permit the use of the fully nonlinear equations of motion.

The Practice of Engineering Dynamics, First Edition. Ronald J. Anderson.
© 2020 John Wiley & Sons Ltd. Published 2020 by John Wiley & Sons Ltd.
Companion Website: www.wiley.com/go/anderson/engineeringdynamics

where $\{F\}$ is a vector of constant force amplitudes and ω is a known *forcing frequency*. It is important that we represent these forces as being harmonic at the forcing frequency and that we use the complex exponential form of the harmonic function. Using only $\sin \omega t$ or $\cos \omega t$ works if there is no damping in the system but the complex exponential form must be used in the general case where damping and, as a result, phase angles are present.

The applied forces replace the zero vector on the right hand side of Equation 7.1 to give,

$$[M]\{\ddot{x}\} + [C]\{\dot{x}\} + [K]\{x\} = \{F\}e^{i\,\omega t}. \tag{7.3}$$

As is for any set of linear differential equations, there are two components of the solution to Equation 7.3. The first is the complementary solution to the homogeneous differential equations (the same equations but with a zero vector on the right hand side). This solution involves the motion of the system at its natural frequencies in response to specified initial conditions. The second is the particular solution that is an oscillation at the forcing frequency. The frequency domain analysis always assumes that the complementary solution will decay with time because of damping in the system and that the particular solution will give the steady state motions in response to the externally applied forces with which we are concerned.

The particular solution is a time-varying vector $\{x(t)\}$ which, along with its first two derivatives, must vary with time as $e^{i\,\omega t}$. The result can only be that $\{x(t)\} = \{X\}e^{i\,\omega t}$ where $\{X\}$ is a vector of complex constants representing the amplitudes and phases of the degrees of freedom in response to the forcing function. Notice that the response will be at the forcing frequency and not at any of the natural frequencies of the system, as determined from the eigenvalue analysis, unless the forcing frequency is equal to a natural frequency.

We substitute

$$\{x(t)\} = \{X\}e^{i\,\omega t}$$
$$\{\dot{x}(t)\} = i\,\omega\{X\}e^{i\,\omega t}$$
$$\{\ddot{x}(t)\} = (i\,\omega)^2\{X\}e^{i\,\omega t} = -\omega^2\{X\}e^{i\,\omega t} \tag{7.4}$$

into Equation 7.3 to get,

$$(-\omega^2[M] + i\,\omega[C] + [K])\{X\}e^{i\,\omega t} = \{F\}e^{i\,\omega t}. \tag{7.5}$$

The $e^{i\,\omega t}$ terms in Equation 7.5 cancel out and a single, complex, coefficient matrix

$$[A(\omega)] = (-\omega^2[M] + i\,\omega[C] + [K]) \tag{7.6}$$

can be defined to represent the group of matrices on the left hand side.

We are left with a simple set of linear algebraic equations

$$[A(\omega)]\{X\} = \{F\}. \tag{7.7}$$

Equation 7.7 is remarkably like the general form of the system of eigenvalue equations presented as Equation 5.11 in Section 5.1 as,

$$[A(\lambda)]\{X\} = \{0\} \tag{7.8}$$

where $[A(\lambda)] = (\lambda^2[M] + \lambda[C] + [K])$ and characteristic values of $\lambda = \sigma + i\,\omega$, the eigenvalues, made the determinant of $[A(\lambda)]$ equal to zero.

The difference is that, in the frequency response equations (Equation 7.7), the coefficient matrix, $[A(\omega)]$, is a function only of an imaginary variable, i ω. As a consequence, the determinant of $[A(\omega)]$ will never be zero as ω is varied since i $\omega \neq \sigma + i\,\omega$ in the general case[2].

Given a forcing frequency ω and a forcing vector $\{F\}$, all of the terms in Equation 7.7 will be known except for $\{X\}$. The problem then becomes one of solving a relatively simple set of linear algebraic equations[3] to find the response amplitudes.

Conceptually, although not computationally because of its inefficiency, it is often useful to look at the solution of sets of linear algebraic equations that would result from an application of Cramer's Rule[4].

Cramer's rule says that, given the set of N linear algebraic equations $[C]\{x\} = \{r\}, x_j$, the jth element of the vector of unknowns $\{x\}$, can be found from the relationship,

$$x_j = \frac{\det[\{C\}_1\{C\}_2 \cdots \{C\}_{j-1}\{r\}\{C\}_{j+1} \cdots \{C\}_N]}{\det[C]} \tag{7.9}$$

where $\{C\}_i$ is the ith column of $[C]$. That is, the right hand side of the set of equations is substituted for the jth column of $[C]$ and x_j is found from the ratio of two determinants.

Applying this to Equation 7.7 in an attempt to find X_j yields,

$$X_j = \frac{\det[\{A(\omega)\}_1\{A(\omega)\}_2 \cdots \{A(\omega)\}_{j-1}\{F\}\{A(\omega)\}_{j+1} \cdots \{A(\omega)\}_N]}{\det[A(\omega)]}. \tag{7.10}$$

The point of this is that, whatever value the numerator of Equation 7.10 might have, the amplitude of any X_j depends strongly on the denominator of the equation, which will become very nearly zero when the forcing frequency, ω, is near a natural frequency of the system. In fact, $\det[A(\omega)]$ is equal to zero when the system is undamped and the forcing frequency is equal to a natural frequency. This situation results in the prediction of infinite response amplitudes when an undamped system is forced at one of its natural frequencies.

Consider, as an example, the system shown in Figure 7.1. The equations of motion are,

$$\begin{bmatrix} m_1 & 0 \\ 0 & m_2 \end{bmatrix} \begin{Bmatrix} \ddot{x}_1 \\ \ddot{x}_2 \end{Bmatrix} + \begin{bmatrix} c_1 + c_2 & -c_2 \\ -c_2 & c_2 \end{bmatrix} \begin{Bmatrix} \dot{x}_1 \\ \dot{x}_2 \end{Bmatrix}$$
$$+ \begin{bmatrix} k_1 + k_2 & -k_2 \\ -k_2 & k_2 \end{bmatrix} \begin{Bmatrix} x_1 \\ x_2 \end{Bmatrix} = \begin{Bmatrix} F_1(t) \\ F_2(t) \end{Bmatrix}. \tag{7.11}$$

If we let,

$$\begin{Bmatrix} F_1(t) \\ F_2(t) \end{Bmatrix} = \begin{Bmatrix} F_1 \\ F_2 \end{Bmatrix} e^{i\,\omega t} \tag{7.12}$$

---

2 There are, of course, cases without damping where $\sigma = 0$ and i ω may indeed be an eigenvalue of the system causing the determinant to be zero. Such cases will be discussed later.
3 The equations are a set of linear algebraic equations with *complex* coefficients. The standard methods for solving linear algebraic equations still apply but must be used with complex arithmetic.
4 Gabriel Cramer (1704–1752), a Swiss mathematician, published, in 1750, a treatise entitled *Introduction á l'analyse des lignes courbes algébraique* where, in a chapter on the classification of curves, he presents the now famous "Cramer's rule".

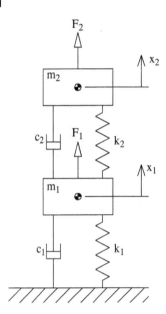

Figure 7.1 A linear system with two masses.

we get, after substitution and canceling of the $e^{i \omega t}$ terms,

$$
\begin{bmatrix}
-\omega^2 m_1 + i \, \omega(c_1 + c_2) + (k_1 + k_2) & -i \, \omega c_2 - k_2 \\
-i \, \omega c_2 - k_2 & -\omega^2 m_2 + i \, \omega c_2 + k_2
\end{bmatrix}
\begin{Bmatrix} X_1 \\ X_2 \end{Bmatrix}
= \begin{Bmatrix} F_1 \\ F_2 \end{Bmatrix}.
$$

(7.13)

After rearranging the elements of the coefficient matrix, this becomes,

$$
\begin{bmatrix}
(-\omega^2 m_1 + k_1 + k_2) + i \, \omega(c_1 + c_2) & -k_2 - i \, \omega c_2 \\
-k_2 - i \, \omega c_2 & (-\omega^2 m_2 + k_2) + i \, \omega c_2
\end{bmatrix}
\begin{Bmatrix} X_1 \\ X_2 \end{Bmatrix}
= \begin{Bmatrix} F_1 \\ F_2 \end{Bmatrix}
$$

(7.14)

which is in the standard form $[A(\omega)]\{X\} = \{F\}$ and the complex coefficients are obvious. The determinant of the coefficient matrix is,

$$
\det[A(\omega)] = [(-\omega^2 m_1 + k_1 + k_2) + i \, \omega \, (c_1 + c_2)][(-\omega^2 m_2 + k_2) + i \, \omega c_2]
$$
$$
- [-k_2 - i \, \omega c_2]^2
$$

(7.15)

and we can use Cramer's rule to solve for X_1 and X_2. The solution is,

$$
X_1 = \frac{F_1[(-\omega^2 m_2 + k_2) + i \, \omega c_2] - F_2[-k_2 - i \, \omega c_2]}{[(-\omega^2 m_1 + k_1 + k_2) + i \, \omega(c_1 + c_2)][(-\omega^2 m_2 + k_2) + i \, \omega c_2] - [-k_2 - i \, \omega c_2]^2}
$$

(7.16)

and,

$$
X_2 = \frac{F_2[(-\omega^2 m_1 + k_1 + k_2) + i \, \omega(c_1 + c_2)] - F_1[-k_2 - i \, \omega c_2]}{[(-\omega^2 m_1 + k_1 + k_2) + i \, \omega(c_1 + c_2)][(-\omega^2 m_2 + k_2) + i \, \omega c_2] - [-k_2 - i \, \omega c_2]^2}.
$$

(7.17)

There is no elegant form of these relationships. The point in showing them is to indicate that solutions can be found and that they are complex. The solutions, once determined, can be written as,

$$X_1 = a_1 + ib_1; \quad X_2 = a_2 + ib_2 \tag{7.18}$$

or,

$$X_1 = \sqrt{a_1^2 + b_1^2} \; \angle\tan^{-1}(b_1/a_1); \quad X_2 = \sqrt{a_2^2 + b_2^2} \; \angle\tan^{-1}(b_2/a_2) \tag{7.19}$$

where a_1, b_1, a_2 and b_2 represent the numerical results of substituting parameter values into the solutions.

Consider the case where the system has the parameter values given in Table 7.1. With these values, an eigenvalue analysis gives two modes as follows.

- Mode 1: 0.797 Hz, 9.33% damping,

$$\begin{Bmatrix} X_1 \\ X_2 \end{Bmatrix} = \begin{Bmatrix} 0.6695 \; \angle 4.9° \\ 1.0000 \; \angle 0° \end{Bmatrix}.$$

- Mode 2: 2.713 Hz, 13.97% damping,

$$\begin{Bmatrix} X_1 \\ X_2 \end{Bmatrix} = \begin{Bmatrix} 1.0000 \; \angle 0° \\ 0.3321 \; \angle -163.0° \end{Bmatrix}.$$

We then expect that if the system is forced at varying frequencies, we should see relatively large response at forcing frequencies near the two natural frequencies: 0.797 Hz and 2.713 Hz. Figures 7.2–7.5 show the forced responses, both amplitudes and phases, as the forces shown in Table 7.1 are applied at varying frequencies.

Mass 1 (the lower mass in Figure 7.1) shows increased response at both of the natural frequencies and has larger amplitude at the first natural frequency. Mass two is quite responsive at the first natural frequency but shows a fairly weak response at the second

Table 7.1 Parameters for the linear system with two masses.

Parameter	Value	Units
m_1	10	kg
m_2	20	kg
c_1	50	N s m^{-1}
c_2	5	N s m^{-1}
k_1	1000	N m^{-1}
k_2	1500	N m^{-1}
F_1	100	N
F_2	10	N

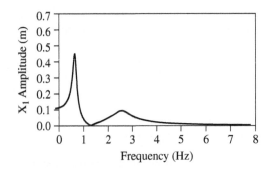

Figure 7.2 Amplitude response of mass 1.

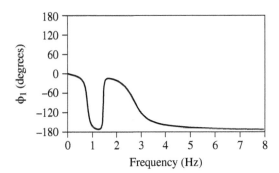

Figure 7.3 Phase response of mass 1.

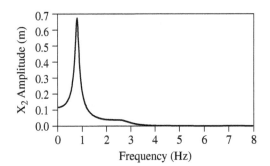

Figure 7.4 Amplitude response of mass 2.

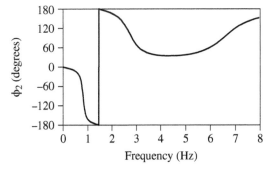

Figure 7.5 Phase response of mass 2.

frequency. Notice on Figure 7.2 that there is a frequency between 1 and 2 Hz where the amplitude of the lower mass is reduced to zero[5].

The phase angle plots show that the two masses are approximately in-phase with each other up to the point between 1 and 2 Hz where the lower mass stops moving. They both have negative phase angles with respect to a positively applied force up to that point. At higher frequencies, the upper mass moves to a positive phase angle and the lower mass maintains a negative phase so that the two masses are approximately out-of-phase with each other.

The observed phase relationship is very much in line with the predicted mode shapes – in-phase at the lower natural frequency and out-of-phase at the higher.

The relative amplitudes at the two natural frequencies are also in line with the mode shapes. Both masses have relatively large amplitudes in the mode shape corresponding to the lower natural frequency with X_2 being larger than X_1. Figures 7.2 and 7.4 show this behavior clearly. The eigenvector at the higher natural frequency indicates that X_1 should be significantly larger than X_2 and the figures indicate that this is so.

In fact, a close look at the predicted response at the natural frequencies shows very close agreement with the mode shapes. The comparison is:

- Mode 1: 0.797 Hz, 9.33% damping,

$$\left\{ \begin{array}{c} X_1 \\ X_2 \end{array} \right\} = \left\{ \begin{array}{c} 0.6695 \angle 4.9° \\ 1.0000 \angle 0° \end{array} \right\}.$$

- Response at 0.80 Hz,

$$\left\{ \begin{array}{c} X_1 \\ X_2 \end{array} \right\} = \left\{ \begin{array}{c} 0.67 \angle = 0.37° \\ 1.0 \angle 0° \end{array} \right\}.$$

- Mode 2: 2.713 Hz, 13.97% damping,

$$\left\{ \begin{array}{c} X_1 \\ X_2 \end{array} \right\} = \left\{ \begin{array}{c} 1.0000 \angle 0° \\ 0.3321 \angle - 163.0°. \end{array} \right\}.$$

- Response at 2.71 Hz,

$$\left\{ \begin{array}{c} X_1 \\ X_2 \end{array} \right\} = \left\{ \begin{array}{c} 1.0 \angle 0° \\ 0.36 \angle - 188° \end{array} \right\}.$$

It can be expected that any system with an externally applied harmonic force that is near a natural frequency, will respond by moving approximately in the mode shape associated with that frequency regardless of where the force is applied to the system.

---

5 The fact that the forced response of a body to which a harmonic force is applied can be reduced to zero simply by having another mass elastically attached to it lies behind the theory of the well known and widely used *vibration absorber*. For undamped systems, the frequency at which the motion of m_1 is reduced to zero is precisely the natural frequency of the system attached to m_1. That is, at the frequency $\sqrt{k_2/m_2}$, the upper mass and spring system will have all of the response and the lower mass will stop moving. In the lightly damped case presented here, the motion of m_1 is minimized at 1.45 Hz where $\sqrt{k_2/m_2} = 1.38$ Hz.

7.2 Seismic Disturbances

Seismic disturbances are harmonic motions of the ground or, more particularly, the supporting base of a system. We consider again the system we considered in Section 7.1 but with the applied forces removed and replaced by a harmonic ground disturbance $y(t)$ given by

$$y(t) = Ye^{i\,\omega t} \tag{7.20}$$

where Y is the amplitude of the base motion and ω is the forcing frequency (see Figure 7.6). The equations of motion are slightly modified from those of Equation 7.9 to become

$$
\begin{bmatrix} m_1 & 0 \\ 0 & m_2 \end{bmatrix} \begin{Bmatrix} \ddot{x}_1 \\ \ddot{x}_2 \end{Bmatrix} + \begin{bmatrix} c_1 + c_2 & -c_2 \\ -c_2 & c_2 \end{bmatrix} \begin{Bmatrix} \dot{x}_1 \\ \dot{x}_2 \end{Bmatrix}
$$
$$
+ \begin{bmatrix} k_1 + k_2 & -k_2 \\ -k_2 & k_2 \end{bmatrix} \begin{Bmatrix} x_1 \\ x_2 \end{Bmatrix} = \begin{Bmatrix} c_1\dot{y} + k_1 y \\ 0 \end{Bmatrix} \tag{7.21}
$$

where we notice that the effect of the ground disturbance is to apply forces only to the lower mass and that these forces pass through the elements connecting that mass to the base.

Since the disturbance is harmonic and at a known frequency, the response will also be harmonic at the same frequency as pointed out in Equation 7.4. That is,

$$\{X(t)\} = \begin{Bmatrix} X_1 \\ X_2 \end{Bmatrix} e^{i\,\omega t}. \tag{7.22}$$

After substitution and simplification, the resulting algebraic equations are

$$
\begin{bmatrix} -\omega^2 m_1 + i\,\omega(c_1 + c_2) + (k_1 + k_2) & -i\,\omega c_2 - k_2 \\ -i\,\omega c_2 - k_2 & -\omega^2 m_2 + i\,\omega c_2 + k_2 \end{bmatrix} \begin{Bmatrix} X_1 \\ X_2 \end{Bmatrix}
$$
$$
= \begin{Bmatrix} i\,\omega c_1 + k_1 \\ 0 \end{Bmatrix} Y. \tag{7.23}
$$

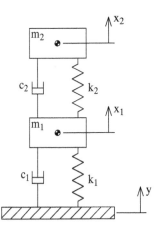

Figure 7.6 Seismic disturbance of a system.

Cramer's rule yields

$$
X_1 = \frac{\det \begin{bmatrix} i\,\omega c_1 + k_1 & -i\,\omega c_2 - k_2 \\ 0 & -\omega^2 m_2 + i\,\omega c_2 + k_2 \end{bmatrix}}{\det \begin{bmatrix} -\omega^2 m_1 + i\,\omega(c_1 + c_2) + (k_1 + k_2) & -i\,\omega c_2 - k_2 \\ -i\,\omega c_2 - k_2 & -\omega^2 m_2 + i\,\omega c_2 + k_2 \end{bmatrix}} Y
\tag{7.24}
$$

and

$$
X_2 = \frac{\det \begin{bmatrix} -\omega^2 m_1 + i\,\omega(c_1 + c_2) + (k_1 + k_2) & i\,\omega c_1 + k_1 \\ -i\,\omega c_2 - k_2 & 0 \end{bmatrix}}{\det \begin{bmatrix} -\omega^2 m_1 + i\,\omega(c_1 + c_2) + (k_1 + k_2) & -i\,\omega c_2 - k_2 \\ -i\,\omega c_2 - k_2 & -\omega^2 m_2 + i\,\omega c_2 + k_2 \end{bmatrix}} Y.
\tag{7.25}
$$

Equations 7.24 and 7.25 give the output response, X_1 and X_2, as a function of the magnitude of the input, Y, and the forcing frequency, ω. These equations can be written as,

$$
X_1 = T_1(\omega)Y
\tag{7.26}
$$

and

$$
X_2 = T_2(\omega)Y
\tag{7.27}
$$

where $T_1(\omega)$ and $T_2(\omega)$ are the complex *transfer functions* relating output, X_1 and X_2, to input magnitude, Y. The complex numbers carry both amplitude and phase information. Often we simply want to know the amplitude relationship between output and input. In that case we take the absolute values and write,

$$
|X_1| = |T_1(\omega)||Y|
\tag{7.28}
$$

and

$$
|X_2| = |T_2(\omega)||Y|.
\tag{7.29}
$$

7.3 Power Spectral Density

Consider the apparently random variable with zero mean value, $x(t)$, shown in Figure 7.7. Assume that we have sampled it and used the discrete Fourier transform[6] to write it in the form of Equation 9.3. That is,

$$
x(t) = a_0 + \sum_{n=1}^{N} a_n \cos(n\omega_0 t) + \sum_{n=1}^{N} b_n \sin(n\omega_0 t).
\tag{7.30}
$$

The term a_0 represents the mean value of $x(t)$. We can drop a_0 in the following since we are considering a variable with a mean value of zero.

---

6 Discrete Fourier transforms are covered in Chapter 9 Analysis of Experimental Data.

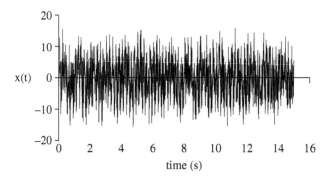

Figure 7.7 A measured variable $x(t)$ plotted versus time.

Equation 7.30 can be rewritten in terms of amplitudes and phase angles [7] as,

$$x(t) = \sum_{n=1}^{N} A_n \cos(n\omega_0 t + \phi_n).$$ (7.31)

We now recognize that the random time signal has been decomposed into a finite number of terms, each of which is a harmonic function with known amplitude and frequency of the form,

$$f(t) = A \cos(\omega t + \phi).$$ (7.32)

The mean-square value of $f(t)$ is,

$$\overline{f}^2 = \frac{1}{T} \int_0^T x^2(t)\, dt$$ (7.33)

where $T = 2\pi/\omega$ is the period of the harmonic function.

Since $f(t)$ repeats cyclically, the mean-square value of any cycle will be equal to the total mean-square value of the function. The total mean-square value can therefore be calculated from a single cycle of the wave and, since the phase angle ϕ does not affect the value, it can be assumed to be zero (note that this is the same as shifting the integration limits to coincide with the beginning and end of a single cycle of the wave).

A cycle is therefore represented by the range $0 \le \omega t \le 2\pi$ resulting in limits on the integration of $0 \le t \le (2\pi/\omega)$. As a result, using Equation 7.32, we can write,

$$\overline{f}^2 = \frac{1}{(2\pi/\omega)} \int_{t=0}^{t=2\pi/\omega} A^2 \cos^2 \omega t\, dt.$$ (7.34)

We define a dummy variable $u = \omega t$ and transform the integration by replacing t with u/ω and dt with du/ω to get,

$$\overline{f}^2 = \frac{A^2}{2\pi} \int_{u=0}^{u=2\pi} \cos^2 u\, du.$$ (7.35)

The integral can be looked up in standard tables of integrals or worked out using integration by parts to yield,

$$\int_{u=0}^{u=2\pi} \cos^2 u\, du = \left[\frac{1}{2}u + \frac{1}{4}\sin 2u \right]_{u=0}^{u=2\pi} = \pi.$$ (7.36)

---

7 See Equation 9.53

Substituting Equation 7.36 into Equation 7.35 gives the total mean-square value of a harmonic function as,

$$\bar{f}^2 = \frac{A^2}{2}. \tag{7.37}$$

Applying Equation 7.37 to each of the terms in Equation 7.31 and adding them together gives the following expression for the total mean-square value of the signal.

$$\bar{x}_T^2 = \sum_{n=1}^{N} \frac{A_n^2}{2}. \tag{7.38}$$

Figure 7.8 shows an example of a Fourier analysis of a signal that has generated five magnitude components. The frequency increment referred to in the figure is f_0[8] whereas it is more convenient to write our equations in terms of ω_0 in rad s^{-1}. The five components of the mean-square value are ($A_n^2/2$ for $\omega = n\omega_0$ where $n = 1, 5$) and are shown in Figure 7.9.

Figure 7.10 shows how the components of the mean-square value can be added together to arrive at the *cumulative mean-square curve*. At each increment of ω_0, the new component of the mean-square value is added to the value that has been accumulated from previous

Figure 7.8 DFT component amplitudes.

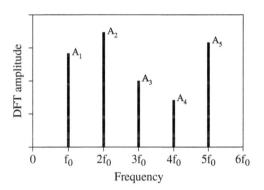

Figure 7.9 Component mean-square values.

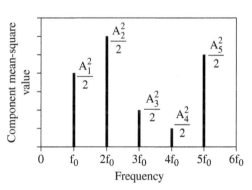

---

8 You will see in Chapter 9 that, if we have a continuous time domain function $x(t)$ and we wish to approximate it in the range $0 \leq t \leq T$ by using a series of harmonic functions, we choose a base frequency f_0 that allows one complete cycle in the range and then add higher frequency components that are multiples of f_0. Clearly, f_0 in cycles per second will be such that there is one cycle in T seconds so that $f_0 = 1/T$ cycles per second or Hz. The base frequency in radians per second will then be $\omega_0 = 2\pi/T$.

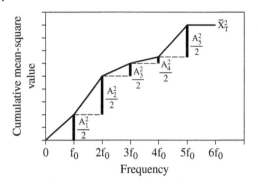

Figure 7.10 The cumulative mean-square value.

components. Eventually the cumulative mean-square curve arrives at the total mean-square value for the signal $\bar{x}_T^2$ as specified in Equation 7.38.

We know the total mean-square value of the signal from the time domain analysis (see Equation 9.2)and the value derived from Equation 7.38 must agree with that value. This fact shows that the terms in the summation (i.e. the values of $A_n^2/2$) must be strong functions of the frequency increment ω_0 that was used to generate the discrete Fourier transform in Equation 7.31. If we use a large value of ω_0 we would expect a small value of N would cover the frequency range of interest and a few relatively large values of A_n would combine to give $\bar{x}_T^2$. Conversely, a small value of ω_0 would require a large value of N to cover the same frequency range and we would see many small values of A_n, again combining to yield $\bar{x}_T^2$.

We consider the contribution that any term makes to the summation in Equation 7.38 as an incremental contribution to the total mean-square value and denote it by $\Delta \bar{x}_n^2$. That is,

$$\Delta \bar{x}_n^2 = \frac{A_n^2}{2}. \tag{7.39}$$

The incremental frequency between terms has been specified to be ω_0. If we consider the cumulative mean-square curve in Figure 7.10, it is clear that, at any given frequency $\omega = n\omega_0$, the slope of the cumulative mean-square curve can be approximated by the ratio $\Delta \bar{x}_n^2/\omega_0$.

We designate this ratio to be $W(\omega)$ and define it to be the *power spectral density* or *PSD*.

As the frequency increment approaches zero, the cumulative mean-square curve approaches a smooth function and the PSD can be defined to be the derivative of the function[9].

$$W(\omega) = \lim_{\omega_0 \to 0} \frac{\Delta \bar{x}^2}{\omega_0} = \frac{d\bar{x}^2}{d\omega}. \tag{7.40}$$

We can once again refer to the component mean-square values shown in Figure 7.9 and to the cumulative mean-square curve of Figure 7.10 to see how they can be interpreted.

---

9 The precise definition of the derivative is,

$$\lim_{\Delta\omega \to 0} \frac{\Delta \bar{x}^2}{\Delta\omega} = \frac{d\bar{x}^2}{d\omega}$$

but, since it is always the case in this analysis that $\Delta\omega = \omega_0$, the notation of Equation 7.40 is acceptable.

Figure 7.11 Cumulative mean-square curve for the example system.

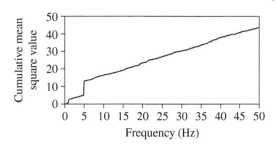

First of all, the component mean-square values are representative of the level of the signal that occurs at a known discrete frequency. That is, if there is a large amount of activity at a frequency (e.g. $2\omega_0$ in Figure 7.9) then the bar will be higher than it is at a frequency without much content (e.g. at $4\omega_0$). The corresponding frequencies in Figure 7.10 show a steep slope or high PSD value approaching $2\omega_0$ and a gentler slope or lower PSD value approaching $4\omega_0$. The result is that the PSD curve will show peaks at frequencies where the content is high (e.g. the natural frequencies of a system).

Figure 7.11 shows the cumulative mean-square curve for the apparently random signal presented in Figure 7.7. Notice that the curve demonstrates a relatively linear growth with frequency over the entire range except for two rapid changes below 10 Hz.

Consider first the linear behavior. If we imagine forming a PSD by taking the derivative or slope of this function, we will end up with a small positive value which is nearly constant. This is indicative of nearly the same amplitude at all frequencies. This is exactly the characteristic of a random signal and the slope indicates that there is a randomness to the signal in Figure 7.7.

The frequencies where the cumulative mean-square value changes rapidly are frequencies where the slope (i.e. the PSD) will increase sharply and then decrease just as rapidly to the previous value. These are peaks in the PSD curve and indicate frequencies where the signal has a great deal of content. These could be natural frequencies of the system or they could be forcing frequencies. Whichever is the case, they are invisible in Figure 9.1 but are of great interest to the analyst.

Figure 7.12 shows the PSD for the apparently random signal being considered. It is shown on a frequency range compressed from that of Figure 7.11 because nothing of interest happens after 10 Hz. The waviness seen near the frequency axis is representative of the less than perfectly random background signal. The two peaks are clearly visible at 1 Hz and 5 Hz.

Figure 7.12 PSD for the example system.

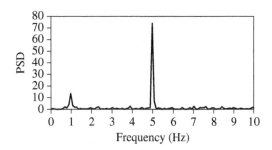

To this point we have demonstrated that it is possible to use the PSD to extract frequency information about a signal. Of course we are also interested in amplitude information. If, for example, the function $x(t)$ being considered in this example is an acceleration measured on a body then we know at which frequencies the accelerations are large. How large are they?

To answer this question we turn to the commonly used *RMS* or *root-mean-square* value of the signal. The name means exactly what it says, the root-mean-square is the square root of the mean-square value of the signal. That is, for a single frequency component such as that presented in Equation 7.32 (i.e. $f(t) = A\sin(\omega t + \phi)$), the mean-square value (Equation 7.37) is $\overline{f}^2 = \frac{A^2}{2}$ and the RMS value is,

$$f_{\text{RMS}} = \sqrt{\overline{f}^2} = \sqrt{\frac{A^2}{2}} = \frac{A}{\sqrt{2}}. \tag{7.41}$$

Equation 7.41, while being commonly quoted, is actually not very useful if you have more than a single frequency. In our case, we have performed a Fourier analysis and have a very large number of frequency components. The RMS value we use will depend on the *bandwidth* in which we are interested. That is, we need to decide on a frequency interval in which to specify the RMS value of the signal. So-called *wideband* RMS covers the entire spectrum of the signal and is a request for the square root of the total mean-square value of the signal. Clearly this gives no frequency discrimination to the amplitude determination. In a case such as the example we are considering we are likely to be more interested in a *narrow band* RMS analysis that gives the RMS value as a function of frequency on a bandwidth of, say, 1 Hz.

The RMS is calculated by integrating under the PSD curve between two specified frequencies, ω_L and ω_R, and then taking the square root of the result. To see this, consider Equation 7.40. It can be rearranged to give,

$$d\overline{x}^2 = W(\omega)d\omega. \tag{7.42}$$

This can be integrated to yield,

$$\int_{\overline{x}_L^2}^{\overline{x}_R^2} d\overline{x}^2 = \int_{\omega_L}^{\omega_R} W(\omega)d\omega \tag{7.43}$$

or,

$$\Delta\overline{x}^2_{\omega_L \to \omega_R} = \int_{\omega_L}^{\omega_R} W(\omega)d\omega \tag{7.44}$$

and the RMS value between ω_L and ω_R is the square root of $\Delta\overline{x}^2_{\omega_L \to \omega_R}$ from Equation 7.44. That is,

$$\text{RMS}_{\omega_L \to \omega_R} = \sqrt{\int_{\omega_L}^{\omega_R} W(\omega)d\omega}. \tag{7.45}$$

7.3.1 Units of the PSD

The units of PSD's often seem arcane and random but they are in fact specified in a consistent manner. Notice that the integral being considered in Equations 7.44 and 7.45 is the area under the PSD curve between the frequency limits ω_L and ω_R. This integral must therefore have units derived from those on the vertical axis multiplied by those on the horizontal or frequency axis. In addition, the units of the area under the PSD curve must satisfy the requirement that the square root of the units give an amplitude of the measured signal. That is, the square root must result in an RMS amplitude.

Consider the case where the signal we have measured is taken from an accelerometer and has been calibrated to specify the acceleration in g's where a g is the acceleration due to gravity. The measured acceleration has been transformed to the frequency domain using an FFT with the frequency specified in Hertz[10] (Hertz are abbreviated as Hz where 1 Hz = 1 cycle s^{-1}).

In this case, the units of the PSD must be g^2/Hz. The area under the curve between two frequencies then has units of $(g^2/\text{Hz}) \times (\text{Hz})$ or g^2. The square root of the area is then $\sqrt{g^2}$ or g which is the desired amplitude unit.

There are many units for the PSD, all of them derived from the units used to measure the quantities characterizing the system in question. For example, the surface roughness of a road is best characterized by measuring the vertical irregularities of the surface and then transforming them, not to the frequency domain, but as a function of wavelength. Road irregularities are used as input disturbances to dynamic models of vehicles and the frequency of the disturbance depends on the wavelength of the irregularity and the speed of the vehicle. In the case of the US customary units[11], the amplitudes of the irregularities will typically be in inches (abbreviated in) and the wavelengths will be in feet/cycle (abbreviated ft/cycle). The units of the PSD will then be $\text{in}^2 \cdot \text{cycle/ft}$. If we repeated the analysis in the SI system, the irregularities would be measured in meters (m) and the wavelengths in meters/cycle (m/cycle) resulting in a PSD in $\text{m}^2 \cdot \text{cycle/m}$. Note the unfortunate coincidence that this can be simplified to m·cycle and the feeling for what dimensions a PSD must have is lost.

7.3.2 Simulation using the PSD

There are instances where the input to a dynamic system cannot be expressed as a known function of time. In these cases, the $F(t)$ input that we use in our models cannot be exactly specified even though we know it must exist. Consider for example, a vehicle driving over uneven ground. The tires follow the profile of the ground and the chassis responds to the up-and-down motion of the tires through the action of the suspension. The forcing input in this case will come from the time varying normal forces acting between the ground and the tires. These forces will be functions of the speed, V, of the vehicle coupled with the profile over which the vehicle is being driven. If, for example, the ground has a purely sinusoidal profile with amplitude, Y, and wavelength, λ, each tire will experience a seismic disturbance[12] exactly like that described in Section 7.2.

---

10 Heinrich Hertz (1857–1894), a German physicist, proved the existence of the electromagnetic waves predicted theoretically by Maxwell. In recognition this, the unit of frequency, one cycle per second, is named the Hertz.

11 US customary units are still widely used in the United States and cannot be dismissed as being irrelevant. (See Appendix C)

12 The amplitude of the seismic input will be Y and the forcing frequency, ω, can be calculated from the period of the input. That is, the vehicle will travel over one wavelength of the disturbance in the period, T,

It is very unlikely that the profile of the ground will ever be a simple sinusoid but, using the Fourier series again, we can fit the random looking profile with a series of sinusoidal functions that, when superimposed, model it very closely. Then, given a vehicle speed, we can calculate the response to each of the frequencies that the vehicle encounters while traveling over the profile. The superposition of all of these responses will then give the total vehicle response.

A typical problem of this type is addressed by firstly obtaining a PSD of the input disturbance. These inputs are usually expressed as PSD amplitude as a function of wavelength (or, sometimes, as a function of one over the wavelength). You can develop them by measuring the profile of your specific input and producing a PSD or you can easily find published PSDs for various systems such as rails, roads, sea states, etc.

The simulation procedure here is as follows: 1. Derive the linearized equations of motion about a stable equilibrium state using degrees of freedom including those that will act as the seismic inputs. 2. Assume harmonic variation of the seismic inputs with known amplitude and frequency. 3. Assume that all DOFs move with the same frequency as the input but have different amplitudes and derive the transfer functions for the degrees of freedom that are not seismic inputs. 4. Calculate the PSDs of the output variables of interest. 5. Integrate the output PSDs over frequency bandwidths of interest and calculate the RMS values for the bandwidths.

The transfer functions we use are of the type we derived in Equations 7.26 and 7.27 in Section 7.2. That is, the output, X_{out}, is related to the input, Y_{in}, by,

$$X_{\text{out}} = T\left(\omega\right) Y_{\text{in}}. \tag{7.46}$$

Since the PSD represents amplitude only, we take the absolute values of the terms in Equation 7.46, and then square both sides of the equation, divide each side by $2\omega_0$ and take the limit as ω_0 goes to zero and find,

$$\left(\lim_{\omega_0 \to 0} \frac{|X_{\text{out}}|^2}{2\omega_0}\right) = |T\left(\omega\right)|^2 \left(\lim_{\omega_0 \to 0} \frac{|Y_{\text{in}}|^2}{2\omega_0}\right). \tag{7.47}$$

Comparing this result to Equation 7.40 shows that the PSD of the output appears on the left hand side and the PSD of the input appears on the right hand side so we can write,

$$PSD_{\text{out}} = |T\left(\omega\right)|^2 PSD_{\text{in}}. \tag{7.48}$$

As an example, consider the case where we are looking at the response of a vehicle as it travels over a terrain for which we know the PSD as a function of wavelength, λ. The first step is to derive the equations of motion, linearized about an equilibrium state where the vehicle is sitting on level ground. The wheels of the vehicle, N_w in number, are given vertical degrees of freedom, $\{x_w\}$. The remainder of the vehicle is given N_v degrees of freedom. These are a set of translations and rotations that the analyst decides are able to describe fully the motion of the vehicle relative to the wheels. They are contained in the vector $\{x_v\}$.

---

giving $VT = \lambda$ and the frequency will be such that $\omega T = 2\pi$. Equating T from each of these expressions gives a forcing frequency $\omega = 2\pi V/\lambda$.

The equations of motion, in partitioned form[13], are,

$$
\begin{bmatrix} \mathbf{M}_{vv} & \mathbf{M}_{vw} \\ \mathbf{M}_{wv} & \mathbf{M}_{ww} \end{bmatrix} \begin{Bmatrix} \ddot{\mathbf{x}}_v \\ \ddot{\mathbf{x}}_w \end{Bmatrix} + \begin{bmatrix} \mathbf{C}_{vv} & \mathbf{C}_{vw} \\ \mathbf{C}_{wv} & \mathbf{C}_{ww} \end{bmatrix} \begin{Bmatrix} \dot{\mathbf{x}}_v \\ \dot{\mathbf{x}}_w \end{Bmatrix}
$$
$$
+ \begin{bmatrix} \mathbf{K}_{vv} & \mathbf{K}_{vw} \\ \mathbf{K}_{wv} & \mathbf{K}_{ww} \end{bmatrix} \begin{Bmatrix} \mathbf{x}_v \\ \mathbf{x}_w \end{Bmatrix} = \begin{Bmatrix} \mathbf{0} \\ \mathbf{F}_w \end{Bmatrix}. \tag{7.49}
$$

The partitioning is shown by the dashed lines in Equation 7.49. The two sub-matrices on the diagonal are square matrices. The two off-diagonal matrices are rectangular. Consider the stiffness matrix as an example. The overall stiffness matrix is square and is dimensioned $(N_v + N_w) \times (N_v + N_w)$. The sub-matrix $\mathbf{K}_{vv}$ is $N_v \times N_v$. $\mathbf{K}_{ww}$ is $N_w \times N_w$. The upper right sub-matrix, $\mathbf{K}_{vw}$, is $N_v \times N_w$ and, on the lower left, $\mathbf{K}_{wv}$ is $N_w \times N_v$.

The right hand side of the equations of motion contains the externally applied forces acting on the system. Since we are considering small motions about equilibrium, we see, in this vector, only those forces that change as we move away from equilibrium. This is the reason for the first N_v elements being zero. The only external forces acting on the vehicle degrees of freedom are gravitational forces and, since they are constant, they are canceled out by constant suspension forces. The wheel degrees of freedom see external forces, $\{\mathbf{F}_w\}$. These are forces of interaction with the ground that vary as the vehicle traverses the profile. The total external force acting on each wheel will be the static normal load in equilibrium plus the dynamic force found from the solution of the equations of motion.

The partitioned equations of motion can be written as two separate sets of differential equations. The first is,

$$
[\mathbf{M}_{vv}]\{\ddot{\mathbf{x}}_v\} + [\mathbf{C}_{vv}]\{\dot{\mathbf{x}}_v\} + [\mathbf{K}_{vv}]\{\mathbf{x}_v\}
$$
$$
= -[\mathbf{M}_{vw}]\{\ddot{\mathbf{x}}_w\} - [\mathbf{C}_{vw}]\{\dot{\mathbf{x}}_w\} - [\mathbf{K}_{vw}]\{\mathbf{x}_w\} \tag{7.50}
$$

where the wheel degrees of freedom have been taken to the right hand side because they represent the seismic forcing input.

The second set is,

$$
[\mathbf{M}_{wv}]\{\ddot{\mathbf{x}}_v\} + [\mathbf{M}_{ww}]\{\ddot{\mathbf{x}}_w\}
$$
$$
+ [\mathbf{C}_{wv}]\{\dot{\mathbf{x}}_v\} + [\mathbf{C}_{ww}]\{\dot{\mathbf{x}}_w\}
$$
$$
+ [\mathbf{K}_{wv}]\{\mathbf{x}_v\} + [\mathbf{K}_{ww}]\{\mathbf{x}_w\} = \{\mathbf{F}_w\}. \tag{7.51}
$$

Equation 7.51 is useful for finding the dynamic loads that the wheels experience as the profile is traversed. This is excellent structural design information but is not always the focus of these simulations. At the early stages of design, engineers are often more interested in the vehicle ride quality, a quantity that can be assessed from the accelerations of the vehicle found by solving Equation 7.50.

---

13 In this example, the wheel degrees of freedom appear as the final N_w elements of the vector containing the degrees of freedom. They don't necessarily have to be ordered like this but the analyst has to be able to keep track of which degrees of freedom they are because, as you will see, they are treated differently. Keeping track of the wheel DOFs is an important bookkeeping exercise.

The total solution will come from the superposition of solutions to harmonic inputs. We have the input PSD as a function of wavelength and we generate solutions for a large number of wavelengths covering the input data. Given a constant forward speed, V, for the vehicle and choosing a wavelength, λ, we can calculate the forcing frequency, $\omega = 2\pi V/\lambda$ (see footnote 12), and write the wheel vertical displacement as,

$$\{x_w\} = A\{e^{i(\omega t+\phi_w)}\} = A\{e^{i\phi_w}\}e^{i\omega t} \tag{7.52}$$

where A is the amplitude of the wave and ϕ_w is a phase angle that accounts for the distance between wheels. That is, if two wheels are spaced exactly a multiple of one wavelength apart, their input will be exactly in phase and the wheels will go up and down together. If we take the leading wheel to have zero phase, then a wheel a distance, ℓ_w, behind it will have a phase angle, $\phi_w = -(\ell_w/\lambda)(2\pi)$. The vector, $\{e^{i\phi_w}\}$, is a constant vector of length, N_w.

Differentiating gives the wheel velocities,

$$\{\dot{x}_w\} = i\omega A\{e^{i\phi_w}\}e^{i\omega t} \tag{7.53}$$

and, differentiating again, gives the wheel accelerations,

$$\{\ddot{x}_w\} = -\omega^2 A\{e^{i\phi_w}\}e^{i\omega t}. \tag{7.54}$$

The steady-state response of the vehicle degrees of freedom will be harmonic at the forcing frequency, giving,

$$\{x_v\} = \{X_v\}e^{i\omega t}. \tag{7.55}$$

Differentiating Equation 7.55 twice and substituting into the equations of motion yields,

$$[-\omega^2 M_{vv} + i\omega C_{vv} + K_{vv}]\{X_v\}e^{i\omega t} =$$
$$- [-\omega^2 M_{vw} + i\omega C_{vw} + K_{vw}]A\{e^{i\phi_w}\}e^{i\omega t}. \tag{7.56}$$

Finally, dividing throughout by the input amplitude, A, and solving gives the transfer functions for all the degrees of freedom of the vehicle (see Equations 7.26 and 7.27). They appear in a vector, $\{T(\omega)\}$, which is defined by,

$$\{T(\omega)\} = \left(\frac{1}{A}\right)\{X_v\}$$
$$= -[-\omega^2 M_{vv} + i\omega C_{vv} + K_{vv}]^{-1}[-\omega^2 M_{vw} + i\omega C_{vw} + K_{vw}]\{e^{i\phi_w}\}. \tag{7.57}$$

We are often interested in predicting accelerations rather than displacements. In this case, the same procedure applies except that the transfer functions used will be those associated with acceleration rather than displacement. Given our assumption of harmonic motion, where acceleration is always $-\omega^2$ multiplied by displacement, we can use acceleration transfer functions,

$$\{T_{acc}(\omega)\} = -\omega^2\{T(\omega)\}. \tag{7.58}$$

Furthermore, given our linearity assumptions, we can create transfer functions for points away from where degrees of freedom are defined simply by producing an expression for the displacement of a point using the defined degrees of freedom and the dimensions that locate the point of interest with respect to the points where degrees of freedom are defined. This is completely analogous to defining linear spring displacements as we have done so many times before.

Once we have the desired transfer function, we simply use Equation 7.48 to calculate the output PSD for a range of frequencies of our choice. The RMS values of the output can be found by integrating the output PSD over specified bandwidths.

Exercises

Descriptions of the systems referred to in the exercises are contained in Appendix A.

7.1 Consider system 3 for the case where the parameters for the vehicle are: mass = 30.0 kg; pitch moment of inertia about G =20.0 kg m^2; a =0.5 m; b =1.0 m; k =3000 N m^{-1}.

Find the transfer functions for vertical motion, x, and pitch rotation, θ, as the vehicle travels at $V = 10.0$ m s^{-1} over sinusoidal terrain having a wavelength $\lambda = 1.0$ m.

7.2 The power spectral density curve for a measured signal, covering the 90 to 110 Hz frequency band, can be approximated by the quadratic,

$$\text{PSD} = -f^2 + 200f - 9800$$

where the PSD has units of $(\text{m s}^{-2})^2 \text{ Hz}^{-1}$ and f is the frequency in Hz.

Calculate the RMS value of this signal for the 90 to 110 Hz frequency band. Using the units of the calculated RMS value, explain what type of signal was originally measured.

7.3 Consider system 10 (see Exercise 3.7 for the equations of motion) for the case where there is, initially, no applied force and the system is in equilibrium with both x and θ equal to zero. A harmonic force, $F(t) = F \cos \omega t$, is then applied to the rod in the direction shown in the figure.

a) Show that the steady state response of the system will be,

$$\begin{Bmatrix} X \\ \Theta \end{Bmatrix} \cos \omega t$$

where,

$$X = \frac{kd^2 F}{(-m\omega^2 + 2k)(-md^2\omega^2 + kd^2) - (-kd)^2}$$

and,

$$\Theta = \frac{Fd(-m\omega^2 + 2k)}{(-m\omega^2 + 2k)(-md^2\omega^2 + kd^2) - (-kd)^2}$$

b) At what forcing frequency, ω, does the amplitude of θ go to zero? This is an example of the *vibration absorber* described in footnote 5.

7.4 The transit vehicle of system 19 runs on a track with a surface displacement PSD that can be described by

$$PSD = \frac{(2 \times 10^{-10})\left(\frac{1}{\lambda}\right)^6}{1 + (1 \times 10^{-2})\left(\frac{1}{\lambda}\right)^3 + (1 \times 10^{-7})\left(\frac{1}{\lambda}\right)^6}$$

where λ is the wavelength in m and the PSD is in $m^2/(cycles/m)$. Use the parameter values from Exercise 6.5(b) and 6.5(d), augmented with a wheel mass of $m_w = 1000$ kg, to perform the following analyses.

a) Extend the model of Exercise 6.5 by adding vertical degrees of freedom to the wheels to get a four degree of freedom model.

b) Use the methods of Subsection 7.3.2 to derive, numerically, the transfer function for vertical acceleration of the center of mass. Use a vehicle speed of $V = 50$ km h^{-1} and a rail surface wavelength of $\lambda = 0.50$ m.

c) Assessment of ride quality in vehicles is often based on RMS accelerations plotted versus 1/3 octave frequency bands[14]. To predict ride quality for the vehicle being considered here would require a dynamic model of the full vehicle and we don't have that. We can, however, use the simple model we have to predict the RMS accelerations of the center of mass of the truck frame. Make that prediction for 1/3 octave bands with center frequencies spanning the range from 1 Hz to 80 Hz. Use a vehicle speed of 50 km h^{-1}.

Note: This will require that you integrate the PSDs over the bandwidths in order to get the RMS accelerations. This will have to be done numerically. At a minimum, you can calculate the PSDs at the bandwidth boundaries and use the trapazoidal rule. To get better accuracy, you will need to calculate the PSDs at more frequencies and use a better numerical integration method.

---

14 An *octave* is defined as a doubling of frequency. That is, starting from a center frequency, f_o, the octaves are $2 \times f_o = 2f_o$, $2 \times 2 \times f_o = 4f_o$, and so on. For any given center frequency, f_c, the resulting *bandwidth* spans the frequencies from $\left[2^{\left(-\frac{1}{2}\right)} \times f_c \text{ to } 2^{\left(\frac{1}{2}\right)} \times f_c\right]$. In dynamics and vibrations, it is common to work with *1/3 octave bands* for which the center frequencies are defined as f_o, $2^{\left(\frac{1}{3}\right)} \times f_o$, $2^{\left(\frac{1}{3}\right)} \times 2^{\left(\frac{1}{3}\right)} \times f_o$, etc. The resulting bandwidth around center frequency, f_c, is $\left[2^{\left(-\frac{1}{6}\right)} \times f_c \text{ to } 2^{\left(\frac{1}{6}\right)} \times f_c\right]$.

8

Time Domain Solutions

The only solution technique that can include all of the nonlinearities in the equations of motion is one which solves the governing differential equations in the time domain. That is, powerful numerical integration techniques work directly with a first-order differential equation representation of the equations of motion, including all of their nonlinear terms, and solve them with respect to time.

The solutions generated in this way are known as *simulations* of the system. They are approximations to what would come from experimental measurements on the real system. There are a few issues to keep in mind when performing simulations.

- The approximation to reality is no better than the data used in the simulation. Detailed dynamic models with very accurate parameter values provide the best simulations.
- The data generated from a time domain simulation have as much value in design as an experiment does. You can see how a system behaves but no natural understanding of the underlying mechanics arises from the simulation. Compare this to a linear stability analysis where the mode shapes give the analyst a "feeling" for what is happening and therefore some idea of how to solve the problem. If you don't like the dynamic behavior of a system you are simulating, you will likely have to resort to a trial and error variation of parameters in order to improve the response.
- One advantage of simulations in software is that it is much cheaper than building and testing prototypes. A good simulation will lead to early design iterations that avoid much of what typically goes wrong with early prototypes. Any system that goes to market will need prototypes built and tested at some stage but the total number required can be minimized by using a good simulation as a starting point.
- Test results can be used to validate simulations so that the designers know they can rely on the simulations to make accurate predictions of how a system's dynamic response will change when system parameters are varied away from their nominal values. In the final design stages, simulations become powerful design tools.

The Practice of Engineering Dynamics, First Edition. Ronald J. Anderson.
© 2020 John Wiley & Sons Ltd. Published 2020 by John Wiley & Sons Ltd.
Companion Website: www.wiley.com/go/anderson/engineeringdynamics

8.1 Getting the Equations of Motion Ready for Time Domain Simulation

The set of N simultaneous nonlinear differential equations of motion, Equation 3.88, from the end of Chapter 3 is the starting point for time domain simulations. It is,

$$[M(q, \dot{q}, t)]\{\ddot{q}\} = \{f(q, \dot{q}, t)\}. \tag{8.1}$$

It is only in the case of time domain simulations that we are able to keep all of the nonlinear terms and use the equations of motion without any simplifying assumptions such as the small motions that allow a linearization of the equations. The solutions are generated by solving the set of ordinary differential equations, Equation 8.1, with respect to time. The solution requires a specified set of initial conditions for both displacement and velocity of all the degrees of freedom.

Standard digital computer solution methods (commonly, although erroneously, referred to as "integrators") require that the differential equations first be put in first-order form. This is completely analogous to the standard eigenvalue form discussed in Section 5.1.

Equation 8.1 can be transformed into an equivalent first-order set as follows.

- Define a new set of N variables for the displacements

$$\{x_1\} = \{q\}. \tag{8.2}$$

- Define a second set of new variables for the rates of change of the displacements

$$\{x_2\} = \{\dot{q}\}. \tag{8.3}$$

- From the preceding two definitions, a new set of N first-order differential equations has been created. These are,

$$\{\dot{x}_1\} = \{x_2\}. \tag{8.4}$$

- With straightforward substitutions, the nonlinear differential equations given in Equation 8.1 can be written in terms of the new variables as,

$$[M(x_1, x_2, t)]\{\dot{x}_2\} = \{f(x_1, x_2, t)\} \tag{8.5}$$

or, solving for $\{\dot{x}_2\}$,

$$\{\dot{x}_2\} = [M(x_1, x_2, t)]^{-1}\{f(x_1, x_2, t)\}. \tag{8.6}$$

- Equations 8.4 and 8.6 can then be written together as a set of $2N$ first-order differential equations,

$$\left\{ \begin{array}{c} \dot{x}_1 \\ \dot{x}_2 \end{array} \right\} = \left\{ \begin{array}{c} x_2 \\ [M(x_1, x_2, t)]^{-1}\{f(x_1, x_2, t)\} \end{array} \right\} \tag{8.7}$$

with user specified initial conditions,

$$\left\{ \begin{array}{c} x_1(0) \\ x_2(0) \end{array} \right\} = \left\{ \begin{array}{c} q(0) \\ \dot{q}(0) \end{array} \right\} = \left\{ \begin{array}{c} \text{displacements at } t = 0 \\ \text{velocities at } t = 0 \end{array} \right\}.$$

Considering Equation 8.7, it is clear that the derivatives (i.e. rates of change) of the displacements ($\{\dot{x}_1\}$) and of the velocities ($\{\dot{x}_2\}$) can be calculated at any time t if the numerical values of the displacements ($\{x_1\}$) and the velocities $\{x_2\}$ are known at that time. That is, only the displacements and velocities appear on the right hand side of Equation 8.7 so that the rates of change on the left hand side of the equations can be easily calculated. Notice, in particular, that specifying the initial conditions allows the calculation of the rates of change at time $t = 0$.

8.2 A Time Domain Example

We first look at an example of the type of solution of the equations of motion that we are seeking here and later discuss the numerical methods used to generate the solution.

Consider the simple pendulum shown in Figure 8.1. It is stationary (i.e. has zero velocity) and is hanging vertically downward at time $t = 0$. A horizontal force, $F(t)$, is suddenly applied to the pendulum bob at time $t = 0$ and is maintained at a constant magnitude and orientation thereafter. We seek to explore how the motion of the pendulum (measured by the angle θ) varies with time for $t \geq 0$.

The equation of motion for this system is easily generated using Lagrange's Equation as follows.

The kinetic and potential (datum at the connection to ground) energy expressions are,

$$T = \frac{1}{2}m(\ell\dot{\theta})^2 \quad \text{and} \quad U = -mg\ell \cos\theta.$$

Taking the necessary partial derivatives and recognizing that the generalized force associated with the angle θ is,

$$Q_\theta = F(t)\ell \cos\theta$$

leads to the single differential equation of motion,

$$m\ell^2\ddot{\theta} + mg\ell \sin\theta = F(t)\ell \cos\theta.$$

Following the procedure established previously in this section for reorganizing the equations of motion for time domain solutions, we define new variables,

$$x_1 = \theta \quad \text{and} \quad x_2 = \dot{\theta}$$

Figure 8.1 The pendulum.

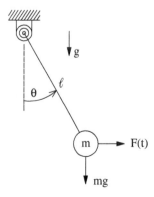

and differentiate them to get,

$$\dot{x}_1 = \dot{\theta} \quad \text{and} \quad \dot{x}_2 = \ddot{\theta}.$$

Then we recognize that $\dot{x}_1$ is identically equal to x_2 as stated in Equation 8.4. This gives a simple first-order differential equation,

$$\dot{x}_1 = x_2.$$

We then substitute the new variables into the differential equation of motion to find,

$$m\ell^2 \dot{x}_2 + mg\ell \sin x_1 = F(t)\ell \cos x_1$$

from which we can solve for $\dot{x}_2$ as,

$$\dot{x}_2 = -\frac{g}{\ell} \sin x_1 + \frac{F(t)}{m\ell} \cos x_1.$$

The two first-order differential equations can then be written together as,

$$\left\{ \begin{array}{c} \dot{x}_1 \\ \dot{x}_2 \end{array} \right\} = \left\{ \begin{array}{c} x_2 \\ -\dfrac{g}{\ell} \sin x_1 + \dfrac{F(t)}{m\ell} \cos x_1 \end{array} \right\}$$

with initial conditions where $\theta(0) = 0$ (i.e. hanging straight down) and $\dot{\theta}(0) = 0$ (i.e. motion-less) at $t = 0$. In terms of the new variables, the initial conditions become,

$$\left\{ \begin{array}{c} x_1(0) \\ x_2(0) \end{array} \right\} = \left\{ \begin{array}{c} 0 \\ 0 \end{array} \right\}.$$

After setting up the equations of motion in this form, we proceed to a numerical solution. For the purposes of this example we will let the time varying force, $F(t)$, start at a value equal to the weight of the pendulum bob ($F(0) = mg$ at $t = 0$), then have it ramp down (i.e. decrease linearly) to $F(1) = 0$ at $t = 1$ s, and hold it steady at $F(t) = 0$ thereafter. This can be stated as,

$$F(t) = mg (1 - t) \text{ for } 0 \le t \le 1 \quad \text{and} \quad F(t) = 0 \text{ for } t > 1.$$

The predicted results are shown in Figure 8.2. They were generated using the variable time-step Runge–Kutta–Fehlberg solution routine. The numerical solution routines for ordinary differential equations, such as these equations of motion, are discussed next and then we will return to a description of the procedure for solving the equations.

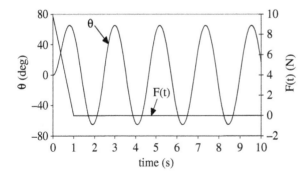

Figure 8.2 Predicted results for the sample time domain simulation.

8.3 Numerical Schemes for Solving the Equations of Motion

Numerical solution schemes for ordinary differential equations use an extrapolation approach that says – *"I know where I am now and I know how quickly things are changing with respect to time so I can estimate where I will be in the future"*. This statement is not mathematical at all but, if you consider it, you will see quickly that the potential for making errors in estimating *"where I will be"* is large. The statement assumes that derivatives remain constant as you extrapolate into the future. If this is true, one can make arbitrarily large steps into the future with zero error.

We can easily show that this is not the case for dynamic systems where there are accelerations. Consider the case where the derivative of the displacement (i.e. the velocity) is constant. Say that $q(t)$ is the displacement and that $\dot{q}(t) = c$ where c is a known constant. Then we can start from an initial displacement $q(0)$ and make predictions about the value of the displacement at any time in the future from the simple expression, $q(t) = q(0) + c\,t$. This is mathematically correct but not applicable to dynamic systems since a constant velocity requires zero acceleration. If we differentiate the original expression, $\dot{q} = c$, we find that, $\ddot{q} = 0$. It immediately becomes clear that we are not working with a system of interest in dynamics since the acceleration is zero.

The fact is, systems with non-zero accelerations do not have solutions that can be extrapolated along straight lines indefinitely. We still rely on extrapolation methods but must be very concerned about the length of time over which we extrapolate because we know that we are trying to follow curved lines even in the simplest case and that we can only extrapolate over small *time steps* without introducing significant errors.

8.4 Euler Integration

The simplest integration scheme is one developed by Leonhard Euler. Extrapolation is often based on the Taylor series expansion of a function and Euler's method is so defined.

We use the following expression for a one-dimensional Taylor series in time

$$x(t_0 + h) = x(t_0) + \left(\frac{dx}{dt}\right)_{t_0} h + \frac{1}{2!}\left(\frac{d^2x}{dt^2}\right)_{t_0} h^2$$

$$+ \frac{1}{3!}\left(\frac{d^3x}{dt^3}\right)_{t_0} h^3 + \cdots + \frac{1}{n!}\left(\frac{d^nx}{dt^n}\right)_{t_0} h^n + \cdots \tag{8.8}$$

where we see that the value of the function $x(t)$ at time $t = t_0 + h$ is equal to the value at $t = t_0$ plus extrapolation terms depending on the derivatives of $x(t)$, all evaluated at time $t = t_0$, and the time step h. The Taylor series expansion of a function is an infinite series where you can see the high order terms going to zero as $n \to \infty$. In most cases, successive terms quickly become negligible in size compared to the sum of the preceding terms before n gets anywhere near ∞. Lagrange developed a theorem that shows that the Taylor series for a function can be truncated at any value of n without loss of accuracy if the nth derivative term is evaluated at a time $t = \tau$ where $t_0 \le \tau \le t_0 + h$. This allows the Taylor series to be written as an exact series with a finite number of terms even though the actual value of τ

cannot be determined,

$$x(t_0 + h) = x(t_0) + \left(\frac{dx}{dt}\right)_{t_0} h + \frac{1}{2!}\left(\frac{d^2x}{dt^2}\right)_{t_0} h^2$$

$$+ \frac{1}{3!}\left(\frac{d^3x}{dt^3}\right)_{t_0} h^3 + \cdots + \frac{1}{n!}\left(\frac{d^nx}{dt^n}\right)_{\tau} h^n. \tag{8.9}$$

In developing integration schemes based on the Taylor series, we will be truncating the series and losing all of the terms after the last one we wish to save. The terms discarded during truncation will introduce an error since they are not zero. This is called the *truncation error* and we wish to keep it as small as possible. Clearly the error will be a function of the time step raised to the power n (i.e. h^n) as can be seen by considering retaining the terms up to and including the h^{n-1} term in Equation 8.9 and discarding the last term. The truncation error introduced by doing this is ϵ_T where,

$$\epsilon_T = \frac{1}{n!}\left(\frac{d^nx}{dt^n}\right)_{\tau} h^n. \tag{8.10}$$

From Equation 8.10, it can be seen that ϵ_T decreases as n increases. The magnitude of ϵ_T is also sensitive to the magnitude of the time step h in that, if h is much less than one, h^n is very small. The magnitude of $\left(\frac{d^nx}{dt^n}\right)_{\tau}$ is finite but indeterminate. Overall, we can control the magnitude of the truncation error for the integration procedure by choosing the values of n and h. The goal is to make the truncation error as small as possible so that the integration scheme approximates the solution to the differential equation very closely.

We now get to the Euler numerical integration scheme. First we define a standard differential equation that we are trying to solve as,

$$\frac{dx}{dt} = f(x, t) \text{ with initial condition } x(0) = x_0. \tag{8.11}$$

Euler suggested truncating the Taylor series immediately after the term that is linear in h. That is, we approximate the function using,

$$x(t_0 + h) = x(t_0) + \left(\frac{dx}{dt}\right)_{t_0} h + \frac{1}{2!}\left(\frac{d^2x}{dt^2}\right)_{\tau} h^2 \tag{8.12}$$

which it has become common to write as,

$$x(t_0 + h) = x(t_0) + \left(\frac{dx}{dt}\right)_{t_0} h + \mathcal{O}(h^2) \tag{8.13}$$

where $\mathcal{O}(h^2)$ shows that the magnitude of the truncation error is *on the order of* h^2 (i.e. the magnitude varies according to the square of the time step).

To use the approximation given in Equation 8.13 to solve the differential equation (Equation 8.11) we need to have an approximation for the first derivative. We can use Equation 8.13 to write,

$$\left(\frac{dx}{dt}\right)_{t_0} = \frac{x(t_0 + h) - x(t_0)}{h} + \mathcal{O}(h) \tag{8.14}$$

where the truncation error on the approximation to the derivative is $\mathcal{O}(h)$ because of the division by h that took place during the reorganization of Equation 8.13 to get Equation 8.14.

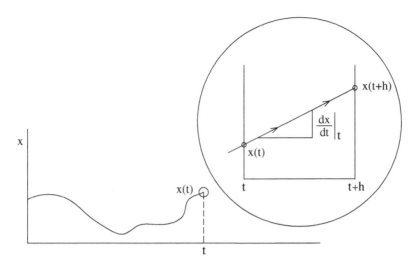

Figure 8.3   Graphical representation of Euler's method.

We now truncate Equation 8.14 and substitute the resulting approximate derivative into the differential equation from Equation 8.11 for $t = t_0$ to get,

$$\frac{x(t_0 + h) - x(t_0)}{h} \approx f(x(t_0), t_0).$$ (8.15)

Rearranging this gives the following extrapolation formula for the Euler integrator

$$x(t_0 + h) \approx x(t_0) + hf(x(t_0), t_0)$$ (8.16)

where we recognize that the truncation error introduced during the extrapolation using this formula is $\mathcal{O}(h^2)$ per time step. If we use the Euler method to propagate a solution from an initial time $(t = 0)$ to a final time $(t = t_f)$, we will need to take N time steps where $t_f = Nh$. From this we can see that $N = t_f/h$ and that the overall truncation error at t_f will be,

$$N \times \mathcal{O}(h^2) = \frac{t_f}{h} \times \mathcal{O}(h^2) = \mathcal{O}(h).$$

We therefore categorize the Euler integrator as an order-h method. Note that the order of the total truncation error is the same as that of the local truncation error on the approximation to the derivative (see Equation 8.14 for the Euler method). We will later discuss integrators with higher order truncation error.

Figure 8.3 gives a graphical explanation for Euler's method. The figure shows that the solution to the differential equation has been propagated from the initial condition out to time t. The derivative, $f(x, t)$, is the local slope, $\frac{dx}{dt}$, at time t. The extrapolation is enlarged in the circle and shows that the method simply "slides up the slope" over the time step.

8.5   An Example Using the Euler Integrator

We consider here a simple differential equation where $f(x, t)$ in Equation 8.11 is simply equal to the function $x(t)$ itself and the initial condition is that $x(0) = 1$. This can be written as,

$$\frac{dx}{dt} = x \text{ with initial condition } x(0) = 1. \tag{8.17}$$

Equation 8.17 is a well-known differential equation and the exact solution is the exponential function. That is,

$$x(t) = e^t. \tag{8.18}$$

We now use the Euler integrator to generate an approximate solution to the differential equation for a time step of h. We will later compare the approximate solution to the exact solution so we can see the effect of the truncation error.

This differential equation is simple enough to allow the approximate solution to be generated in a tabular format (see Table 8.1). We construct the table by first filling in a column with the time in steps of h. Beside this we write the value of $x(t)$ at each time noting that the initial value in this column is given by the initial condition in Equation 8.17 but that subsequent entries in this column arise from extrapolation using the Euler integrator (Equation 8.16). The third column contains the derivative $f(x, t)$ calculated from the differential equation using the values of x and t in the same row of the table. The fourth column contains the approximate value of $x(t + h)$ calculated using the Euler integrator. This value is then carried down to become the next value of $x(t)$ in column two. The process is repeated from there as the approximate solution to the differential equation is propagated forward in time. After a few rows are filled in, the pattern becomes obvious and we can write the approximate solution at the final simulated time $t_s = Nh$ to complete the table.

The exact solution at $t = Nh$ is $x(Nh) = e^{Nh}$ and the approximate solution is $x(Nh) \approx (1 + h)^N$. To see how the two solutions differ we compare their infinite series representations. The exact solution can be written as,

$$x(Nh) = e^{Nh} = \sum_{i=0}^{i=\infty} \frac{(Nh)^i}{i!} = 1 + Nh + \frac{(Nh)^2}{2} + \frac{(Nh)^3}{6} + \cdots . \tag{8.19}$$

Table 8.1 Euler integration example.

t	$x(t)$	$f(x,t)$	$x(t + h)$
0	1	1	$(1 + h)$
h	$(1 + h)$	$(1 + h)$	$(1 + h)^2$
$2h$	$(1 + h)^2$	$(1 + h)^2$	$(1 + h)^3$
$3h$	$(1 + h)^3$	$(1 + h)^3$	$(1 + h)^4$
$4h$	$(1 + h)^4$	$(1 + h)^4$	$(1 + h)^5$
$\vdots$	$\vdots$	$\vdots$	$\vdots$
$t_s = Nh$	$(1 + h)^N$	–	–

The approximate solution $((1 + h)^N)$ can be expanded using the binomial theorem,

$$(a + b)^n = \sum_{k=0}^{k=n} \frac{n!}{(n - k)!k!} a^{n-k}b^k = a^n + na^{n-1}b$$

$$+ \frac{n(n - 1)}{2} a^{n-2}b^2 + \frac{n(n - 1)(n - 2)}{6} a^{n-3}b^3 + \cdots \tag{8.20}$$

where $a = 1, b = h$, and $n = N$ in this case. The result is,

$$x(Nh) \approx (1 + h)^N = 1 + Nh + \frac{N(N - 1)}{2}h^2 + \frac{N(N - 1)(N - 2)}{6}h^3 + \cdots . \tag{8.21}$$

The difference between the exact solution and the approximate solution can be found by subtracting Equation 8.21 from Equation 8.19. This is the truncation error ϵ_T and can be written as,

$$\epsilon_T = \frac{N}{2}h^2 + \frac{N}{6}(3N - 2)h^3 + \cdots . \tag{8.22}$$

Notice that the largest term in the truncation error is $\mathcal{O}(h^2)$. For typical integration time steps of $h = 0.001$ or smaller, the h^2 term is on the order of 1000 times larger than the h^3 term so the truncation error can be approximated as,

$$\epsilon_T \approx \frac{N}{2}h^2 = \frac{(Nh)h}{2} = \frac{t_s h}{2} \tag{8.23}$$

where t_s is the total length of time simulated (see Table 8.1 where $t_s = Nh$). As a result, the truncation error at any simulated time (t_s) is $\mathcal{O}(h)$ and varies linearly with the time-step. Notice that it also varies linearly with the total simulation time as may be expected.

8.6 The Central Difference Method: An $\mathcal{O}(h^2)$ Method

We can develop an integrator with higher order truncation error using the truncated Taylor series in Equation 8.9 where we now keep terms up to $\mathcal{O}(h^3)$,

$$x(t_0 + h) = x(t_0) + \left(\frac{dx}{dt}\right)_{t_0} h + \frac{1}{2!}\left(\frac{d^2x}{dt^2}\right)_{t_0} h^2 + \frac{1}{3!}\left(\frac{d^3x}{dt^3}\right)_{\tau} h^3 \tag{8.24}$$

and a second truncated Taylor series that goes back one time step from t_0. That is,

$$x(t_0 - h) = x(t_0) - \left(\frac{dx}{dt}\right)_{t_0} h + \frac{1}{2!}\left(\frac{d^2x}{dt^2}\right)_{t_0} h^2 - \frac{1}{3!}\left(\frac{d^3x}{dt^3}\right)_{\tau} h^3. \tag{8.25}$$

We subtract Equation 8.25 from Equation 8.24 to get,

$$x(t_0 + h) - x(t_0 - h) = 2h\left(\frac{dx}{dt}\right)_{t_0} + \frac{2h^3}{3!}\left(\frac{d^3x}{dt^3}\right)_{\tau} \tag{8.26}$$

from which we can write the derivative,

$$\left(\frac{dx}{dt}\right)_{t_0} = \frac{x(t_0 + h) - x(t_0 - h)}{2h} - \frac{h^2}{3!}\left(\frac{d^3x}{dt^3}\right)_{\tau} \tag{8.27}$$

or,

$$\left(\frac{dx}{dt}\right)_{t_0} = \frac{x(t_0 + h) - x(t_0 - h)}{2h} + \mathcal{O}(h^2). \tag{8.28}$$

The approximation to the derivative that we use is then,

$$\left(\frac{dx}{dt}\right)_{t_0} \approx \frac{x(t_0 + h) - x(t_0 - h)}{2h} \tag{8.29}$$

where it is recognized that the truncation error is $\mathcal{O}(h^2)$.

This method is called the *central difference method* and it has a higher order truncation error than the Euler integrator and should therefore produce more accurate results for the same time step.

The propagation formula (derived from Equation 8.29 and the fact that $f(x(t_0), t_0)$ is the actual value of the derivative at time t_0) is,

$$x(t_0 + h) = x(t_0 - h) + 2hf(x(t_0), t_0). \tag{8.30}$$

Notice that this method cannot start with only the initial conditions at time $t = 0$. It requires information from one time step previous to the current time and that information is not available at $t = 0$. Methods like this are classified as being *non-self-starting methods*. Another method must be used to propagate the solution out to $t = h$ before the central difference method can be used.

Figure 8.4 shows the magnitude of the errors generated by the Euler integration method and the central difference method as a function of the time step used. The example is the same one used above in Equation 8.17 with the exponential solution as stated in Equation 8.18. The percent errors quoted are those that exist after propagating the solution to $t = 10$ s. According to Equation 8.18, the actual solution at that time is $x(10) = e^{10} = 22026$ and this is used as a basis for calculating the percentage error. The exact solution at $t = h$ and the initial condition at $t = 0$ were used to get the central difference method started. The Euler method started with the initial condition only.

The difference between the two methods is clear with the Euler method beginning to lose accuracy at much smaller time steps than the central difference method. For

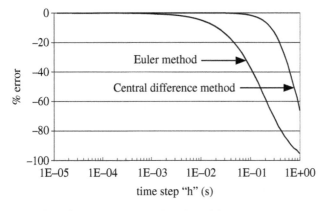

Figure 8.4 Solution errors as a function of time step.

example, at $h \approx 0.02$ the Euler error is nearly -10% and the central difference error is only -0.07%.

8.7 Variable Time Step Methods

To see how variable time step methods work, the *midpoint method*, a self-starting method with $\mathcal{O}(h^2)$ truncation error is introduced. This is a method that uses information from the midpoint of the time step being considered. It will later be combined with the Euler method in order to form a method with variable time steps.

Let there be an ordinary differential equation

$$\frac{dx}{dt} = f[x(t)] \quad \text{with the initial condition } x(0) = x_0 \tag{8.31}$$

with a solution, $x(t)$, that has been propagated to time t_s. The solution at $t_s + h$ is desired. The midpoint method makes the extrapolation in two steps. First the value of $x(t)$ at the midpoint of the step (i.e. $x(t_s + h/2)$) is estimated by an Euler step from t_s using the derivative $f[x(t_s)]$. This midpoint value is designated as x_{mp}.

$$x_{mp} = x(t_s) + \frac{h}{2}f[x(t_s)]. \tag{8.32}$$

The derivative at the midpoint, $f[x_{mp}]$, is then calculated and used to propagate across the entire time step. The estimated solution at time $t_s + h$ is then,

$$x(t_s + h) = x(t_s) + hf[x_{mp}]. \tag{8.33}$$

Figure 8.5 shows the steps graphically.

Notice that this method requires two evaluations of the derivative per time step and uses a point within the time step. This is a feature of many higher order methods that will be discussed later.

The truncation error of this method is determined as follows. First Equation 8.32 is substituted into Equation 8.33 to give,

$$x(t_s + h) = x(t_s) + hf[x(t_s) + \frac{h}{2}f[x(t_s)]]. \tag{8.34}$$

Figure 8.5 The midpoint method.

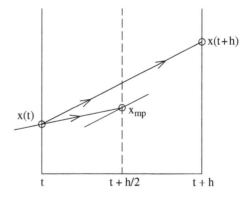

The derivative at the midpoint, $f[x(t_s) + \frac{h}{2}f[x(t_s)]]$, appears in Equation 8.34 and can be expanded using a Taylor series as follows[1],

$$f[x(t_s) + \frac{h}{2}f[x(t_s)]] = f[x(t_s)] + \frac{h}{2}f[x(t_s)]\frac{df[x(t_s)]}{dx} + \mathcal{O}(h^2). \tag{8.35}$$

This is substituted into Equation 8.34 to yield,

$$x(t_s + h) = x(t_s) + hf[x(t_s)] + h\left(\frac{h}{2}\right)f[x(t_s)]\frac{df[x(t_s)]}{dx} + h(\mathcal{O}(h^2)). \tag{8.36}$$

This can be simplified somewhat by noting that (see equation 8.31),

$$f[x(t)] = \frac{dx}{dt} \text{ so that } f[x(t)]\frac{df[x(t)]}{dx} = \frac{dx}{dt}\frac{df[x(t)]}{dx} = \frac{df[x(t)]}{dt} = \frac{d^2x}{dt^2}$$

Substituting this into Equation 8.36 and noting that $h(\mathcal{O}(h^2)) = \mathcal{O}(h^3)$ gives,

$$x(t_s + h) = x(t_s) + h\left(\frac{dx}{dt}\right)_{t_s} + \frac{h^2}{2}\left(\frac{d^2x}{dt^2}\right)_{t_s} + \mathcal{O}(h^3) \tag{8.37}$$

which is the Taylor series expansion for $x(t)$ across the time step with $\mathcal{O}(h^3)$ local truncation error and thus $\mathcal{O}(h^2)$ global truncation error. Based on the global truncation error, the midpoint method is classed as a second-order, self-starting integration technique.

Now consider the case where two different methods, Euler and midpoint, are used simultaneously to find a solution to the differential equation. They differ in truncation error by one order so, given that they use the same time step, it can be expected that they will give slightly different solutions.

The difference between the solutions can be determined from the Taylor series approximations.

$$x_{MP}(t_s + h) = x(t_s) + h\left(\frac{dx}{dt}\right)_{t_s} + \frac{h^2}{2}\left(\frac{d^2x}{dt^2}\right)_{t_s} + \mathcal{O}(h^3) \tag{8.38}$$

and,

$$x_{EU}(t_s + h) = x(t_s) + h\left(\frac{dx}{dt}\right)_{t_s} + \mathcal{O}(h^2). \tag{8.39}$$

Subtracting gives the difference, ϵ, as,

$$\epsilon = x_{MP}(t_s + h) - x_{EU}(t_s + h) = \frac{h^2}{2}\left(\frac{d^2x}{dt^2}\right)_{t_s} + \mathcal{O}(h^3) - \mathcal{O}(h^2) \tag{8.40}$$

which is numerically indeterminate but, assuming a small time step h, is $\mathcal{O}(h^2)$ so that ϵ can be expressed as some constant, A, multiplied by h^2.

$$\epsilon = Ah^2. \tag{8.41}$$

---

1 This step may not be obvious. It can be helpful to first write the Taylor series as,

$$f(x + \Delta x) = f(x) + \left(\frac{df}{dx}\right)_x \Delta x + \mathcal{O}(h^2)$$

and then note that, in this case,

$$\Delta x = \frac{h}{2}f[x(t_s)].$$

The goal is to choose a value of h that keeps the difference between the solutions small. To that end, the user specifies a tolerance on the difference, ϵ_{tol}, that must be satisfied in order to maintain an accurate solution. If the difference between the solutions is larger than the acceptable tolerance, the step-size must be changed in order to bring it under control. Equation 8.41 allows the constant, A, to be determined as,

$$A = \frac{\epsilon}{h^2}.$$

(8.42)

There will be a smaller step size, h_*, that makes the difference between the solutions satisfy the tolerance specification. That step size can be found from,

$$\epsilon_{tol} = Ah_*^2 \text{ which yields } h_*^2 = \frac{\epsilon_{tol}}{A}.$$

(8.43)

Substituting the value of A from Equation 8.42 into Equation 8.43 gives the step size that will satisfy the tolerance as,

$$h_* = \sqrt{\frac{\epsilon_{tol}}{\epsilon}} \, h.$$

(8.44)

Using the variable step size method therefore requires the user to specify a tolerance and an initial step size. The numerical routine can then adjust the step size throughout the solution so that the tolerance is maintained. If the error is too large the step size is reduced and, if the difference between the solutions is much smaller than the tolerance, the step size can be increased to speed up the solution.

This has been a very rudimentary discussion of a complex field of numerical analysis and is only meant to give an impression of the logic behind solution routines that employ variable step size. There are many powerful routines available that employ this method and they are the preferred methods of solution for simulations.

8.8 Methods with Higher Order Truncation Error

Introducing a few simple numerical methods for solving ordinary differential equations has demonstrated the concept of *truncation error* and how it varies with the time step used for the solution. The development of methods with higher order truncation error has been of interest to numerical analysis experts for decades and there are many methods with very high order truncation error ($\mathcal{O}(h^{10})$ and higher methods exist). The truncation errors introduced by these methods will clearly be smaller than those of lower order methods but one must also consider the computational efficiency of the methods. The total error in a numerical solution is not strictly due to truncation. There is also an error introduced by round-off during any calculation. The numbers being used have a finite accuracy and something is lost during each calculation as the numbers are rounded off to match the limited number of significant figures in the computer. As a result, the round-off error grows with the number of computations and the number of computations grows as the time-step gets smaller. There is thus a trade-off between truncation error and round-off error. A very large step-size will result in extreme truncation error and almost no round-off error. A step-size very close to zero will yield almost no truncation error but extreme round-off errors due to the number of calculations required to propagate the solution for any length of time. The result is an "optimum", not too small and not too large, time step that minimizes the total solution error. Unfortunately, it is impossible to know the value of that time step so numerical solutions make use of "reasonable" time steps. Simulations with fixed time steps are often executed twice using different time steps in order to ensure that the solution is not overly sensitive to the time step used.

The most often used higher order method is the classical fourth-order Runge–Kutta[2] formula that requires four derivative ($f(x, t)$) evaluations per step. For the case where t_s is the time at the beginning of a step and x_s is the known value of $x(t_s)$, the derivatives are,

$$k_1 = f(x_s, t_s) \tag{8.45}$$

$$k_2 = f\left(x_s + h\frac{k_1}{2}, t_s + \frac{h}{2}\right)$$

$$k_3 = f\left(x_s + h\frac{k_2}{2}, t_s + \frac{h}{2}\right)$$

$$k_4 = f(x_s + hk_3, t_s + h)$$

and the extrapolation formula is,

$$x_{s+h} = x_s + h\left(\frac{k_1}{6} + \frac{k_2}{3} + \frac{k_3}{3} + \frac{k_4}{6}\right). \tag{8.46}$$

Notice that the derivative is required at the beginning of the step, twice at the midpoint, and finally at the end of the step. Since no information is required previous to the beginning of the step, this is a self-starting method.

Variable time step versions of the Runge–Kutta method are very common and robust. One of the best known is the Runge–Kutta–Fehlberg Method. It simultaneously generates fourth and fifth order solutions using six derivative evaluations as follows.

$$k_1 = f(x_s, t_s) \tag{8.47}$$

$$x_{1/4} = x_s + \frac{h}{4}k_1$$

$$k_2 = f\left(x_{1/4}, t_s + \frac{1}{4}h\right)$$

$$x_{3/8} = x_s + \frac{3}{32}hk_1 + \frac{9}{32}hk_2$$

$$k_3 = f\left(x_{3/8}, t_s + \frac{3}{8}h\right)$$

$$x_{12/13} = x_s + \frac{1932}{2197}hk_1 - \frac{7200}{2197}hk_2 + \frac{7296}{2197}hk_3$$

$$k_4 = f\left(x_{12/13}, t_s + \frac{12}{13}h\right)$$

$$x_1 = x_s + \frac{439}{216}hk_1 - 8hk_2 + \frac{3680}{513}hk_3 - \frac{845}{4104}hk_4$$

---

2 Carl David Tolme Runge (1856–1927) was a German mathematician, physicist, and spectroscopist. Martin Wilhelm Kutta (1867–1944) was a German mathematician. In 1895 Carl Runge published the first Runge–Kutta method. In 1905 Martin Kutta described the popular fourth-order Runge–Kutta method.

$$k_5 = f(x_1, t_s + h)$$

$$x_{1/2} = x_s - \frac{8}{27}hk_1 + 2hk_2 - \frac{3544}{2565}hk_3 + \frac{1859}{4104}hk_4 - \frac{11}{40}hk_5$$

$$k_6 = f\left(x_{1/2}, t_s + \frac{1}{2}h\right).$$

The fourth order Runge–Kutta approximation is then,

$$(x_{s+h})_4 = x_s + \frac{25}{216}hk_1 + \frac{1408}{2565}hk_3 + \frac{2197}{4104}hk_4 - \frac{1}{5}hk_5 \tag{8.48}$$

and the fifth order approximation is,

$$(x_{s+h})_5 = x_s + \frac{16}{135}hk_1 + \frac{6656}{12825}hk_3 + \frac{28561}{56430}hk_4 - \frac{9}{50}hk_5 + \frac{2}{55}hk_6. \tag{8.49}$$

Given a tolerance, ϵ_{tol}, the scaling factor for step size, s, such that the correct step size is $h_* = sh$ is,

$$s = \left(\frac{\epsilon_{tol}h}{2|(x_{s+h})_5 - (x_{s+h})_4|}\right)^{1/4}. \tag{8.50}$$

8.9 The Structure of a Simulation Program

This chapter ends with a description of the way in which a time domain simulation program is structured.

There are three parts to any software designed for simulation of dynamic motions. They will be called here, the main program, the solution routine, and the derivative routine. They are shown graphically in the flowchart of Figure 8.6 and described below.

1. The main program. This is used primarily for input and output functions. Parameter values necessary for defining the system (e.g. masses, moments of inertia, spring constants – anything that is required to numerically form the equations of motion) are accepted as input. These parameter values are stored in common memory so that they can be used by the derivative routine. The main program sets the variables to their initial conditions and then calls the solution routine, sending the current values of the variables and time as well as a time when the updated solution is to be returned. The main program receives the values of the degrees of freedom from the solution routine at the specified time, writes them to an output file, checks to see if the total simulation times has been exceeded, and then returns control to the solution routine or stops.

2. The solution routine. This is an implementation of one of the numerical solution methods described earlier in this chapter. The main program supplies it with numerical values of the variables, the starting time for the step, the time for the next output and the required tolerance (for variable time step methods). The solution routine needs to be able to call the derivative routine to get the derivatives whenever required. Recall that this operation happens four times per step for the classical fourth-order Runge–Kutta method and six times per step for the Runge–Kutta–Fehlberg method. The solution routine retains control and propagates the solution of the differential equations forward in

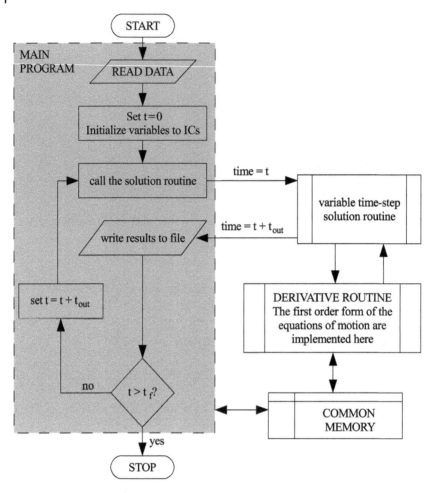

Figure 8.6 Structure of a simulation program.

time until it reaches the time at which the main program is set to write output. It then returns control to the main program.

3. The derivative routine. This is where the equations of motion are coded. This routine is designed simply to provide the derivatives to the solution routine whenever requested. The solution routine calls the derivative routine, providing the current time and the values of the variables, and expects, in return, the numerical values of the derivatives it needs in order to propagate the solution. The derivative routine needs access to the system parameters that the main program stored in common memory in order to set up the equations of motion and solve them for the derivatives.

Consider, as an example, the pendulum from Section 8.2 (see Figure 8.1). The equations of motion are repeated here for easy reference. They are,

$$\begin{Bmatrix} \dot{x}_1 \\ \dot{x}_2 \end{Bmatrix} = \begin{Bmatrix} x_2 \\ -\frac{g}{\ell} \sin x_1 + \frac{F(t)}{m\ell} \cos x_1 \end{Bmatrix}$$

with,

$$F(t) = mg\,(1 - t) \text{ for } 0 \leq t \leq 1 \quad \text{and} \quad F(t) = 0 \text{ for } t > 1.$$

Table 8.2 shows the pseudocode for the Main Program and the Derivative Routine for this system. Any other system would have exactly the same programming structure.

Systems with more than one degree of freedom have derivative routines that are more complicated than the example used in Table 8.2. The general case was presented in Equation 8.7, repeated here as Equation 8.51.

Table 8.2 Pendulum simulation pseudo-code.

MAIN PROGRAM

START

    define common memory for m, g, and ℓ

    open an input file

        read system parameters (m, g, and ℓ)

        read initial conditions ($x_1(0)$ and $x_2(0)$)

        read time between output values (t_{out})

        read total simulation time(t_f)

        read integration control parameters (ϵ, h_{mint}, h_{max}, h_{start})

    WHILE t is less than t_f

    set $t_{ret} = t + t_{out}$

    call the solution routine with the current values of $\{x\}$,

        t, and t_{ret}

    receive updated values of $\{x\}$ at $t = t_{ret}$

    write values of t and $\{x\}$ to the output file

    ENDWHILE

STOP

DERIVATIVE ROUTINE

START

    receive m, g, and ℓ from common memory

    receive $\{x\}$ and t from the solution routine

    IF t is greater than 1

        set $F(t) = 0$

    ELSE

        set $F(t) = mg\,(1 - t)$

    ENDIF

    set $\dot{x}_1 = x_2$

    set $\dot{x}_2 = -\frac{g}{\ell} \sin x_1 + \frac{F(t)}{m\ell} \cos x_1$

RETURN

$$\left\{ \begin{array}{c} \dot{x}_1 \\ \dot{x}_2 \end{array} \right\} = \left\{ \begin{array}{c} x_2 \\ [M(x_1,x_2,t)]^{-1}\{f(x_1,x_2,t)\} \end{array} \right\} \qquad (8.51)$$

with user specified initial conditions

$$\left\{ \begin{array}{c} x_1(0) \\ x_2(0) \end{array} \right\} = \left\{ \begin{array}{c} q(0) \\ \dot{q}(0) \end{array} \right\} = \left\{ \begin{array}{c} \text{displacements at } t = 0 \\ \text{velocities at } t = 0 \end{array} \right\}.$$

It is clear that, in the general case, a set of linear algebraic equations has to be solved whenever the derivative routine responds to a request for the values of the derivatives.

Table 8.3 General simulation pseudo-code.

MAIN PROGRAM

START
    define common memory for system parameters
    open an input file
        read system parameters
        read initial conditions
        read time between output values (t_{out})
        read total simulation time(t_f)
        read integration control parameters $(\epsilon, h_{min}, h_{max}, h_{start})$
    WHILE t is less than t_f
        set $t_{ret} = t + t_{out}$
        call the solution routine with the current values of $\{x\}$,
        receive updated values of $\{x\}$ at $t = t_{ret}$
        write values of t and $\{x\}$ to the output file
    ENDWHILE
STOP

DERIVATIVE ROUTINE

START
    receive system parameters from common memory
    receive $\{x\}$ and t from the solution routine
    set $\{\dot{x}_1\}$ to $\{x_2\}$
    calculate the RHS of the equations of motion $(\{f(x_1,x_2,t)\})$
    calculate the mass matrix $([M(x_1,x_2,t)])$
    send $[M(x_1,x_2,t)]$ and $\{f(x_1,x_2,t)\}$ to a linear algebra solver
    receive $\{\dot{x}_2\}$
RETURN

Equation 8.51 specifies $2N$ derivatives. The first N are obtained by simply equating elements of the vector $\{\dot{x}_1\}$ to $\{x_2\}$ N times. The second N require that Equation 8.5 (repeated here as Equation 8.52 be solved

$$[M(x_1, x_2, t)]\{\dot{x}_2\} = \{f(x_1, x_2, t)\}. \tag{8.52}$$

In the text the solution has been written as if the inverse of the mass matrix had been calculated and then multiplied by the right hand side of the equations of motion. While convenient for use in the text, this is actually a very inefficient way to solve linear algebraic equations. It is much better to use a routine that is designed for efficient solutions of linear algebraic equations such as Gaussian elimination. The pseudocode for the general case is shown in Table 8.3.

Exercises

Descriptions of the systems referred to in the exercises are contained in Appendix A.

8.1 You previously found, in Exercise 3.2, that the equations of motion for system 1 are

$$\begin{bmatrix} (m_1 + m_2)d_1^2 + m_2 d_2(d_2 + 2d_1 \cos\theta_2) & m_2 d_2(d_2 + d_1 \cos\theta_2) \\ m_2 d_2(d_2 + d_1 \cos\theta_2) & m_2 d_2^2 \end{bmatrix} \begin{Bmatrix} \ddot{\theta}_1 \\ \ddot{\theta}_2 \end{Bmatrix}$$
$$= \begin{Bmatrix} m_2 d_1 d_2 \dot{\theta}_2 (2\dot{\theta}_1 + \dot{\theta}_2)\sin\theta_2 - (m_1 + m_2)gd_1\cos\theta_1 - m_2 gd_2\cos(\theta_1 + \theta_2) \\ -m_2 d_1 d_2 \dot{\theta}_1^2 \sin\theta_2 - m_2 gd_2\cos(\theta_1 + \theta_2) \end{Bmatrix}.$$

(a) Set up these equations in the form you would need for a time domain solution.

(b) For the case where $d = 0.05$ m and $m = 1.0$ kg, implement an Euler integration scheme and solve the equations of motion for 20 s of simulated time. Experiment with different time steps to see how the solution is affected.

8.2 The nonlinear equation of motion for system 6 was found twice, in Exercises 2.6 and 3.4. In Exercise 4.3, one equilibrium value of θ was specified to be 60° and the second was found numerically. Exercise 5.4 asked for an analysis of the stability of the two equilibria.

(a) Create a time domain simulation of system 6 using the parameters determined in Exercise 4.3.

(b) Run your simulation for arbitrary initial conditions and note that the system is always trying to get to the stable equilibrium state but cannot settle out there because energy needs to be removed from the system in order for it to stop moving. Undamped, oscillatory solutions about the stable equilibrium value of θ should be clear, especially if you start with no initial velocity and an initial value of θ close to the stable equilibrium value.

(c) Add viscous damping to the equation of motion (a moment on the arm opposing, and proportional to, $\dot{\theta}$ will do the trick) and see if your time domain solution converges to the same stable equilibrium value you found in Exercise 4.3.

8.3 Create a time domain simulation of system 12 using the equation of motion from Exercise 3.12. Use parameter values: $m = 1000$ kg, $k = 30\ 000$ N m^{-1}, and $d = 1.2$ m.

(a) Implement the classical, fourth-order, Runge–Kutta method described in Section 8.8 and use it for the simulation.

(b) Calculate the equilibrium value of θ from the equation of motion using the techniques from Chapter 4. Start your simulation from $\theta = 0$ with zero initial velocity and plot the results versus time. You should see an oscillation about the equilibrium value you found.

(c) Linearize the equation of motion about equilibrium and find the natural frequency analytically using the techniques of Chapter 5. How well does this match the frequency you get by counting cycles on the plotted results of the nonlinear simulation?

8.4 Use parameter values of your choice for system 16 and solve the nonlinear equations of motion (see Exercise 3.9) with initial conditions: $\theta(0) = 90°$ and $\phi(0) = 10°$. Using your predicted values of θ and ϕ, sketch the system at several times over one cycle of motion and visualize the animated motion of rod AB.

Part III

Working with Experimental Data

9

Experimental Data – Frequency Domain Analysis

Up to now, we have concentrated on the theoretical simulation of systems in order to obtain information useful in their design. The theoretical study of dynamic systems and the ability to understand the physical behavior of the systems that the theory imparts are immensely important to practicing engineers. The theory is what leads to the initial system design through simulation and the theory is what gives the ability to understand and correct the behavior of the prototype systems when they are tested.

Equally as important to engineers is the analysis of experimentally derived response data for systems. The validation of the models used in the design phase relies on being able to record and analyze test data. Such analysis is most often done in the frequency domain and that is what we will concentrate on here.

9.1 Typical Test Data

Dynamic measurements are made by sensors of various kinds and are most often recorded in digital form[1]. Figure 9.1 shows a typical measured dynamic variable $x(t)$ plotted as a function of time. Measured variables very often appear to be random because, even after looking at the signal over the entire plot, there is no way to predict where the next point will be.

Characterization of the time domain signal is unrewarding. The mean value of the signal can be found but is not representative of anything but the "average" value of $x(t)$. It contains no information with respect to peak values for instance. Given that $x(t)$ is known in the range $0 \leq t \leq T$, the mean value is defined as

$$\bar{x} = \frac{1}{T} \int_0^T x(t) \, dt. \tag{9.1}$$

---

[1] The analog measurements that used to be recorded on analog devices such as multi-channel tape recorders have been almost completely superseded by in-field A/D converters and digital recording devices. The sampling rate of the A/D converters must be chosen very carefully to satisfy the requirements of the data analysis that will be done later. With analog devices it was possible to re-play the experiments if you got the sampling rate wrong. Now you are required to get it right or repeat the, sometimes very costly, experiments.

The Practice of Engineering Dynamics, First Edition. Ronald J. Anderson.
© 2020 John Wiley & Sons Ltd. Published 2020 by John Wiley & Sons Ltd.
Companion Website: www.wiley.com/go/anderson/engineeringdynamics

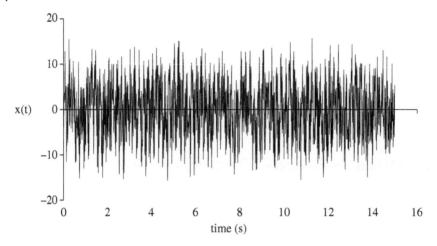

Figure 9.1 A measured variable $x(t)$ plotted versus time.

A characteristic of the signal that can be calculated and that has meaning is the total *mean-square value* of the signal, $\overline{x}_T^2$. The total mean-square value of the signal is defined as the mean value of another signal generated by squaring the original signal and then finding its mean value over the period $0 \leq t \leq T$ using Equation 9.2. Figure 9.2 shows the squared signal and the mean-square value $\overline{x}_T^2 = 43$. Characterizing the signal by this single scalar value seems less than satisfying but that is all that can be done in the time domain. However, the total mean-square value will be useful to us when we discuss scaling of results

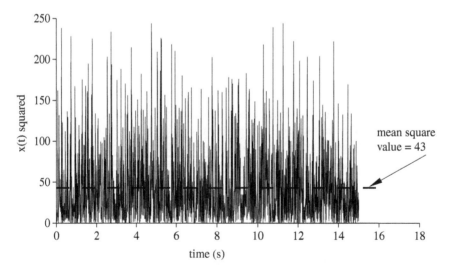

Figure 9.2 The square of $x(t)$ plotted versus time.

in the frequency domain in Section 9.8

$$\bar{x}_{\mathrm{T}}^2 = \frac{1}{T} \int_0^T x^2(t)\, dt \tag{9.2}$$

9.2 Transforming to the Frequency Domain – The CFT

The most common approach to analyzing experimental signals begins with a transformation that takes the measured data into the frequency domain. Readers may be familiar with the terms *FFT* and *fast Fourier transform* that describe a technique that is widely used to accomplish the transformation but may not familiar with the details of the transformation. The mathematical development of the Fourier transform equations is most easily done using complex exponentials and that is the approach used in modern reference books. Unfortunately, the complex exponential approach, while being mathematically elegant, leaves the reader without any physical feeling for what lies behind the equations. The goal here is to present the techniques in a manner that leaves the reader with an understanding of the techniques and a physical feeling for what they represent.

It is best to think of the transformation as a curve fitting exercise. That is, we have a continuous time domain function $x(t)$ and we wish to approximate it in the range $0 \le t \le T$ by using a series of harmonic functions. We choose a base frequency f_0 that allows one complete cycle in the range and then add higher frequency components that are multiples of f_0. Clearly, f_0 in cycles per second will be such that there is one cycle in T seconds so that $f_0 = 1/T$ cycles per second or Hz. The base frequency in radians per second will then be $\omega_0 = 2\pi/T$.

Let the approximating function $x_{\mathrm{A}}(t)$ be expressed in terms of $(2N+1)$ undetermined coefficients a_0, a_n, and b_n as follows

$$x_{\mathrm{A}}(t) = a_0 + \sum_{n=1}^{N} a_n \cos(n\omega_0 t) + \sum_{n=1}^{N} b_n \sin(n\omega_0 t). \tag{9.3}$$

While there are several techniques that could be used to find the "best" values of the coefficients, we will use the traditional least squares[2] approach because it is commonly used in fitting curves to experimental data. We first define an objective function J that is a measure of the difference between the experimental data and the approximating function. Since the difference $[x(t) - x_{\mathrm{A}}(t)]$ can be either positive or negative, we can't use it to establish the goodness of fit so we work with the square of the difference instead and define,

$$J = \frac{1}{2} \int_0^T [x(t) - x_{\mathrm{A}}(t)]^2 dt. \tag{9.4}$$

We choose the coefficients such that they minimize the objective function, thereby minimizing the difference between the experimental curve $x(t)$ and the approximating function

---

2 Appendix D has a description of least squares curve fitting for those who would like to review the method before proceeding with the remainder of this chapter.

$x_A(t)$. The conditions for minimizing J are that the partial derivatives of J with respect to all of the undetermined coefficients must simultaneously be equal to zero. J can be expanded into[3].

$$J = \frac{1}{2} \int_0^T \left[x(t) - a_0 - \sum_{n=1}^N a_n \cos(n\omega_0 t) - \sum_{n=1}^N b_n \sin(n\omega_0 t) \right]^2 dt \tag{9.5}$$

The partial derivative with respect to a_0 is

$$\frac{\partial J}{\partial a_0} = \int_0^T \left[-x(t) + a_0 + \sum_{n=1}^N a_n \cos(n\omega_0 t) + \sum_{n=1}^N b_n \sin(n\omega_0 t) \right] dt = 0. \tag{9.6}$$

For the remaining coefficients,

$$\frac{\partial J}{\partial a_p} = \int_0^T \left[-x(t) + a_0 + \sum_{n=1}^N a_n \cos(n\omega_0 t) + \sum_{n=1}^N b_n \sin(n\omega_0 t) \right] \cos(p\omega_0 t) dt = 0,$$

$$p = 1, N \tag{9.7}$$

and

$$\frac{\partial J}{\partial b_p} = \int_0^T \left[-x(t) + a_0 + \sum_{n=1}^N a_n \cos(n\omega_0 t) + \sum_{n=1}^N b_n \sin(n\omega_0 t) \right] \sin(p\omega_0 t) dt = 0,$$

$$p = 1, N. \tag{9.8}$$

Consider first satisfying the condition on a_0 from Equation 9.6. This can be written as

$$-\int_0^T x(t) dt + a_0 \int_0^T dt + \sum_{n=1}^N a_n \left[\int_0^T \cos(n\omega_0 t) dt \right] + \sum_{n=1}^N b_n \left[\int_0^T \sin(n\omega_0 t) dt \right] = 0. \tag{9.9}$$

Three of the four integrals in Equation 9.9 can be directly evaluated, yielding

$$\int_0^T dt = T; \quad \int_0^T \cos(n\omega_0 t) dt = 0 \; ; \quad \int_0^T \sin(n\omega_0 t) dt = 0. \tag{9.10}$$

The integrals of the trigonometric functions give the area under them as time goes from zero to T. Since they have been defined as multiples of a single wavelength spanning $0 \le t \le T$, they will all complete an integer number of cycles and the area under them will be zero. A mathematical proof is left as an exercise.

The result is that a_0 is

$$a_0 = \frac{1}{T} \int_0^T x(t) dt. \tag{9.11}$$

Comparing this to Equation 9.1, we see that a_0 is the mean value of $x(t)$. Examination of the approximating function in Equation 9.3 makes it clear that this must be the case. The

---

3 You may wonder why a_0 is included and b_0 is not. In fact, a more elegant way of writing the series is

$$x_A(t) = \sum_{n=0}^N a_n \cos(n\omega_0 t) + \sum_{n=0}^N b_n \sin(n\omega_0 t)$$

but it is clear that $\sin(n\omega_0 t) = 0$ when $n = 0$ so the form shown in Equation 9.3 is universal.

function has one constant term (a_0) and superimposes harmonics on top of it. The constant value can only be the mean value.

We now turn to the conditions on the other coefficients as specified in Equations 9.7 and 9.8. These involve integrals of the form

$$\int_0^T \cos(n\omega_0 t)\sin(p\omega_0 t)dt \tag{9.12}$$

$$\int_0^T \sin(n\omega_0 t)\sin(p\omega_0 t)dt \tag{9.13}$$

$$\int_0^T \cos(n\omega_0 t)\cos(p\omega_0 t)dt \tag{9.14}$$

and

$$\int_0^T \sin(n\omega_0 t)\cos(p\omega_0 t)dt \tag{9.15}$$

all of which can be shown to be zero so long as $n \neq p$ (see Exercise 9.2). This is a result of the functions 1, $\cos pt$, and $\sin pt$ being *orthogonal* over the interval $[0 \leq pt \leq 2\pi]$. This concept of orthogonality will reappear when we consider discrete Fourier transforms in Section 9.4.

In the case where $p = n$ Equations 9.7 and 9.8 become

$$-\int_0^T x(t)\cos(n\omega_0 t)dt + a_0\int_0^T \cos(n\omega_0 t)dt + a_n\int_0^T \cos^2(n\omega_0 t)dt$$

$$+b_n\int_0^T \sin(n\omega_0 t)\cos(p\omega_0 t)dt = 0 \tag{9.16}$$

and

$$-\int_0^T x(t)\sin(p\omega_0 t)dt + a_0\int_0^T \sin(p\omega_0 t)dt + a_n\int_0^T \cos(n\omega_0 t)\sin(p\omega_0 t)dt$$

$$+b_n\int_0^T \sin^2(n\omega_0 t)dt = 0. \tag{9.17}$$

Several of the integrals in Equations 9.16 and 9.17 have already been shown to be zero. In addition, it can easily be shown that

$$\int_0^T \cos^2(n\omega_0 t)dt = \int_0^T \sin^2(n\omega_0 t)dt = \frac{T}{2} \tag{9.18}$$

and, as a result, we find that the least squares curve fit to the experimental data is

$$x_A(t) = a_0 + \sum_{n=1}^N a_n\cos(n\omega_0 t) + \sum_{n=1}^N b_n\sin(n\omega_0 t) \tag{9.19}$$

where

$$a_0 = \frac{1}{T}\int_0^T x(t)dt \tag{9.20}$$

$$a_n = \frac{2}{T}\int_0^T x(t)\cos(n\omega_0 t)dt, \ n = 1, N \tag{9.21}$$

$$b_n = \frac{2}{T}\int_0^T x(t)\sin(n\omega_0 t)dt, \ n = 1, N. \tag{9.22}$$

If we now let N increase so that we minimize the objective function at more and more points we see that as N approaches infinity the approximation becomes equal to the function $x(t)$ so that we can write

$$x(t) = a_0 + \sum_{n=1}^{\infty} a_n \cos(n\omega_0 t) + \sum_{n=1}^{\infty} b_n \sin(n\omega_0 t) \tag{9.23}$$

where

$$a_0 = \frac{1}{T} \int_0^T x(t) dt \tag{9.24}$$

$$a_n = \frac{2}{T} \int_0^T x(t) \cos(n\omega_0 t) dt, \quad n = 1, \infty \tag{9.25}$$

$$b_n = \frac{2}{T} \int_0^T x(t) \sin(n\omega_0 t) dt, \quad n = 1, \infty. \tag{9.26}$$

The transformation given by Equations 9.23 through 9.26 is the well-known continuous Fourier rransform (CFT)[4].

Notice that we have considered a continuous function $x(t)$ that was defined over the interval $0 \le t \le T$. The Fourier series representation of this function is based on this interval even though it can be evaluated for any time t. There is an implicit assumption that the function being analyzed is periodic with a period equal to T. This fact will become important to us as we proceed.

9.3 Transforming to the Frequency Domain – The DFT

We have shown that a periodic, continuous function can be exactly represented by a curve fit using an infinite series of harmonic functions, with a base frequency derived from the period of the function and multiples of that frequency. Clearly, practicing engineers will never have to deal with functions like this. Measurements we take are not continuous but are sampled and stored at some sampling rate that gives an equal time between samples, Δt. The measured values are not periodic but we have to finish sampling at some time so we get data over the range $0 \le t \le T$ where T is the time at which we stop sampling. As a result, we get a finite number of points to analyze. Consider the curve fitting process again but without a continuous function to fit this time.

Let there be $2N$ measured data points stored in a vector $x(t)$. Let the sampling rate be $f_s = 1/\Delta t$ where Δt is the constant time between samples. Define the base frequency f_0 in Hz as

$$f_0 = \frac{1}{2N\Delta t} \tag{9.27}$$

giving a base frequency ω_0 in $rad\ s^{-1}$

$$\omega_0 = \frac{2\pi}{2N\Delta t} = \frac{\pi}{N\Delta t}. \tag{9.28}$$

---

4 Jean-Baptiste Joseph Fourier (1768–1830), a French mathematician, is best remembered for his work on the propagation of heat in solid bodies and for his expansions of functions as trigonometric series that we now call Fourier series.

Use a series of harmonics to fit the measured data. The series is

$$x(t) = \sum_{n=0}^{N} [a_n \cos(n\omega_0 t) + b_n \sin(n\omega_0 t)]. \tag{9.29}$$

Since we have $2N$ data points, we can use them to find exactly $2N$ coefficients, a_n and b_n, in Equation 9.29. As it stands, Equation 9.29 has $(2N + 2)$ undetermined coefficients. We can rewrite the series with its first and last terms extracted to see which two coefficients are unnecessary. We do this for some time $t = m\Delta t$ where $0 \leq m \leq (2N - 1)$. Substituting Equation 9.28 into Equation 9.29 and extracting the first and last terms gives $2N$ simultaneous equations of the form

$$x_m = x(m\Delta t) = a_0 \cos(0) + b_0 \sin(0)$$

$$+ \sum_{n=1}^{N-1} \left[a_n \cos\left(\frac{n\pi}{N}m\right) + b_n \sin\left(\frac{n\pi}{N}m\right) \right]$$

$$+ a_N \cos(\pi m) + b_N \sin(\pi m), \ 0 \leq m \leq (2N - 1) \tag{9.30}$$

where we note that $\cos(0) = 1$ and $\sin(0) = \sin(\pi m) = 0$ so that the coefficients b_0 and b_N do not contribute to the curve fit and can be removed[5]. This leaves us with exactly $2N$ undetermined coefficients and $2N$ data points so we can write $2N$ simultaneous equations of the form

$$x_m = a_0 + \sum_{n=1}^{N-1} \left[a_n \cos\left(\frac{n\pi}{N}m\right) + b_n \sin\left(\frac{n\pi}{N}m\right) \right]$$

$$+ a_N \cos(\pi m), \ 0 \leq m \leq (2N - 1). \tag{9.31}$$

The equations can be assembled into the standard matrix form for linear, algebraic equations with a known coefficient matrix, $[C]$, multiplied by a vector of unknowns, $\{a_0 \ldots a_N \ b_1 \ldots b_{N-1}\}^T$, set equal to a vector of known values $\{x_0 \ldots x_{2N-1}\}^T$

$$[C] \left\{ \begin{array}{c} a_0 \\ \vdots \\ a_N \\ b_1 \\ \vdots \\ b_{N-1} \end{array} \right\} = \left\{ \begin{array}{c} x_0 \\ \vdots \\ x_{2N-1} \end{array} \right\} \tag{9.32}$$

where the $2N$ by $2N$ coefficient matrix, $[C]$, has terms c_{ij}, for the a_n (left side of the matrix)

$$c_{ij} = \cos\left[\frac{(i-1)(j-1)\pi}{N} \right] \ ; \ i = 1, 2N, j = 1, N+1 \tag{9.33}$$

and, for the b_n (right side of the matrix)

$$c_{ij} = \sin\left[\frac{(i-1)(j-N-1)\pi}{N} \right] \ ; \ i = 1, 2N, j = N+2, 2N. \tag{9.34}$$

---

5 We are treating the case where we have an even number of samples. The case for an odd number of samples, $(2N + 1)$, can be handled in a similar way with the small difference that an extra coefficient, b_N, is retained.

Solving for the unknown coefficients using Equation 9.32 is a simple matter of using a standard elimination technique such as Gaussian elimination. The computational effort required to do so is approximately proportional to N^3 where "computational effort" is measured in terms of the number of multiplications and divisions required to implement a solution on a digital computer.

9.4 Transforming to the Frequency Domain – A Faster DFT

Here, we return to the concept of least squares curve fitting that we used in Section 9.2 but now it is applied to the finite number of data points we discussed in Section 9.3. We start with the series from Equation 9.31 rewritten so that a_0 and a_N are taken into the summation

$$x_m = \sum_{n=0}^{N} a_n \cos\left(\frac{n\pi}{N}m\right) + \sum_{n=1}^{N-1} b_n \sin\left(\frac{n\pi}{N}m\right) , \quad 0 \le m \le (2N-1) \tag{9.35}$$

and construct an objective function, J, as

$$J = \frac{1}{2} \sum_{m=0}^{2N-1} \left[x_m - \sum_{n=0}^{N} a_n \cos\left(\frac{n\pi}{N}m\right) - \sum_{n=1}^{N-1} b_n \sin\left(\frac{n\pi}{N}m\right)\right]^2 . \tag{9.36}$$

As before, we set the partial derivatives of J with respect to each of the coefficients to zero. That is,

$$\frac{\partial J}{\partial a_p} = -\sum_{m=0}^{2N-1} x_m \cos\left(\frac{p\pi}{N}m\right) + \sum_{m=0}^{2N-1}\left[\sum_{n=0}^{N} a_n \cos\left(\frac{n\pi}{N}m\right)\cos\left(\frac{p\pi}{N}m\right)\right.$$
$$\left. + \sum_{n=1}^{N-1} b_n \sin\left(\frac{n\pi}{N}m\right)\cos\left(\frac{p\pi}{N}m\right)\right] = 0 , \quad p = 0, N \tag{9.37}$$

and

$$\frac{\partial J}{\partial b_p} = -\sum_{m=0}^{2N-1} x_m \sin\left(\frac{p\pi}{N}m\right) + \sum_{m=0}^{2N-1}\left[\sum_{n=0}^{N} a_n \cos\left(\frac{n\pi}{N}m\right)\sin\left(\frac{p\pi}{N}m\right)\right.$$
$$\left. + \sum_{n=1}^{N-1} b_n \sin\left(\frac{n\pi}{N}m\right)\sin\left(\frac{p\pi}{N}m\right)\right] = 0 , \quad p = 1, N-1. \tag{9.38}$$

We used orthogonality in Section 9.2 to show that the vast majority of the integrals involving products of harmonic functions were equal to zero. Orthogonality also holds in the case of summations. The rules are as follows (for $0 \le p, n \le N$),

$$\sum_{m=0}^{2N-1} \sin\left(\frac{n\pi}{N}m\right)\sin\left(\frac{p\pi}{N}m\right) = \begin{cases} 0 & \text{if } p = n = 0, N \\ 0 & \text{if } p \ne n \\ N & \text{if } p = n \ne 0 \end{cases} \tag{9.39}$$

$$\sum_{m=0}^{2N-1} \sin\left(\frac{n\pi}{N}m\right)\cos\left(\frac{p\pi}{N}m\right) = 0 \tag{9.40}$$

$$\sum_{m=0}^{2N-1} \cos\left(\frac{n\pi}{N}m\right)\cos\left(\frac{p\pi}{N}m\right) = \begin{cases} 0 & \text{if } p \neq n \\ N & \text{if } p = n \neq 0, N \\ 2N & \text{if } p = n = 0, N \end{cases}. \tag{9.41}$$

Applying the orthogonality rules to the conditions on the partial derivatives specified in Equations 9.37 and 9.38 yields the following expressions for the coefficients.

$$a_0 = \frac{1}{2N} \sum_{m=0}^{2N-1} x_m \tag{9.42}$$

$$a_n = \frac{1}{N} \sum_{m=0}^{2N-1} x_m \cos\left(\frac{n\pi}{N}m\right), 1 \leq n \leq N-1 \tag{9.43}$$

$$a_N = \frac{1}{2N} \sum_{m=0}^{2N-1} x_m \cos(\pi m) \tag{9.44}$$

$$b_n = \frac{1}{N} \sum_{m=0}^{2N-1} x_m \sin\left(\frac{n\pi}{N}m\right), 1 \leq n \leq N-1. \tag{9.45}$$

The computational effort required to calculate the coefficients using the series given in Equations (9.42–9.45) is approximately proportional to N^2. Comparing this to the N^3 result given in Section 9.3, where a set of linear, algebraic equations was solved, shows the inherent advantage to using the series solution, especially when N is large, as it usually is. For $N = 1000$, using the series method is approximately 1000 times faster than using the linear, algebraic equations.

9.5   Transforming to the Frequency Domain – The FFT

Up to this point, we have been attempting to maintain a geometric feeling for the transformations by describing the process as a curve fitting exercise using sines and cosines to represent the experimentally measured data. From a purely mathematical view, it is much better to do the analysis in the complex domain where the Fourier transform is given by

$$X(f, T) = \int_0^T x(t)e^{-i2\pi ft}dt \text{ where } i = \sqrt{-1}. \tag{9.46}$$

Euler's formula

$$e^{i\theta} = \cos\theta + i\sin\theta \tag{9.47}$$

shows how this is equivalent to the sine and cosine series we have been using.

The *fast Fourier transform* or *FFT* was developed in the mid-1960s. It is beyond the scope of this text to develop the actual algorithm. Suffice it to say that the computational effort is significantly less than other methods. The FFT algorithm requires that the number of data points be a power of two (i.e. $2N = 2^P$ in our cases where we were working with $2N$ points). The computational effort for the FFT is proportional to $N\log_2 N$ which is significantly less than for the series method presented in Section 9.4. For example, if $N = 2^{10} = 1024$, the computational effort of the three methods we have discussed is summarized in Table 9.1.

Table 9.1 Fourier transforms computational effort.

Method	for $N = 2^P$	e.g. $p = 10$	Computer time
Linear algebraic equations	2^{3p}	1.06×10^9	104 858
Series	2^{2p}	1.05×10^6	102
FFT	$p2^p$	1.02×10^4	1

The column labeled *computer time* shows the relative amount of computer processing time that the three methods would use on the same computer and is not in any particular unit of time. The table makes it abundantly clear why the FFT is the method of choice for all analysts. Implementations of the algorithm are commonplace and can even be found in spreadsheets.

9.6 Transforming to the Frequency Domain – An Example

As an example of the techniques for transforming to the frequency domain, consider the function

$$x(t) = 2 + 5\sin(2\pi t) + 10\sin(4\pi t) + 15\cos(6\pi t) \tag{9.48}$$

which is shown in Figure 9.3. This function has a mean value of 2 and three harmonics at frequencies of 1, 2, and 3 Hz. The harmonics have amplitudes of 5, 10, and 15 respectively. Consider sampling the function at a rate of 10 samples per second ($\Delta t = 0.10$ s) and collecting 20 samples ($2N = 20$). The data are shown in Table 9.2. The base frequency for the

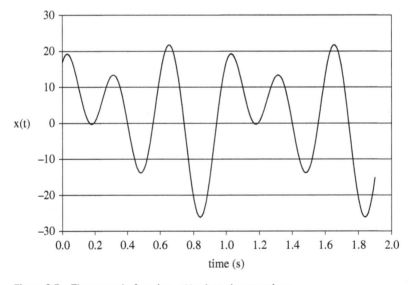

Figure 9.3 The example function, $x(t)$, plotted versus time.

Table 9.2 Sampled data.

n	Time (s)	x(t)
1	0.0	17.000
2	0.1	9.814
3	0.2	0.498
4	0.3	13.013
5	0.4	0.064
6	0.5	−13.000
7	0.6	13.207
8	0.7	15.258
9	0.8	−20.768
10	0.9	−15.085
11	1.0	17.000
12	1.1	9.814
13	1.2	0.498
14	1.3	13.013
15	1.4	0.064
16	1.5	−13.000
17	1.6	13.207
18	1.7	15.258
19	1.8	−20.768
20	1.9	−15.085

transformation is given by Equation 9.27. That is,

$$f_0 = \frac{1}{2N\Delta t} = \frac{1}{20 \times 0.10} = 0.50 \text{ Hz} \tag{9.49}$$

and we expect 20 coefficients, $a_0 \ldots a_{10}$ and $b_1 \ldots b_9$. Solving the set of linear, algebraic equations from Section 9.3 or using the series from Section 9.4 both result in the coefficients shown in Table 9.3. Notice that the FFT can't be used on this data set since the number of points is not a power of two.

The original function can then be reconstructed using the coefficients in Table 9.3 and Equation 9.3. That is, retaining only the non-zero coefficients,

$$x_A(t) = a_0 + a_6 \cos(6\omega_0 t) + b_2 \sin(2\omega_0 t) + b_4 \sin(4\omega_0 t) \tag{9.50}$$

where

$$\omega_0 = 2\pi f_0 = 2\pi \times 0.50 = \pi \tag{9.51}$$

yielding

$$x_A(t) = 2.000 + 15.000 \cos(6\pi t) + 5.000 \sin(2\pi t) + 10.000 \sin(4\pi t) \tag{9.52}$$

which is exactly the same as the original function defined in Equation 9.48.

Table 9.3 DFT coefficients.

Coefficient	Value	Frequency (Hz)
a_0	2.000	0.00
a_1	0.000	0.50
a_2	0.000	1.00
a_3	0.000	1.50
a_4	0.000	2.00
a_5	0.000	2.50
a_6	15.000	3.00
a_7	0.000	3.50
a_8	0.000	4.00
a_9	0.000	4.50
a_{10}	0.000	5.00
b_1	0.000	0.50
b_2	5.000	1.00
b_3	0.000	1.50
b_4	10.000	2.00
b_5	0.000	2.50
b_6	0.000	3.00
b_7	0.000	3.50
b_8	0.000	4.00
b_9	0.000	4.50

So it is clear that the transformation to the frequency domain can be undone and we can transform back to the time domain. This is called the *inverse Fourier transform*.

There is another way to write the DFT. We can use amplitudes and phase angles instead of the coefficients, a_n and b_n. We did this in Section 6.1 when we were considering eigenvectors. We can write

$$a_n \cos(n\omega_0 t) + b_n \sin(n\omega_0 t) = A_n \cos(n\omega_0 t + \phi_n) \tag{9.53}$$

where A_n is the amplitude

$$A_n = \sqrt{a_n^2 + b_n^2} \tag{9.54}$$

and ϕ_n is the phase angle

$$\phi_n = \arctan(b_n/a_n). \tag{9.55}$$

In fact, most data analysis concentrates on amplitudes only and phase angles are ignored. The function $x(t)$ that we have been considering (Equation 9.48 or 9.53), would appear on a plot of amplitude versus frequency as shown in Figure 9.4. We will be working strictly with amplitudes in the following.

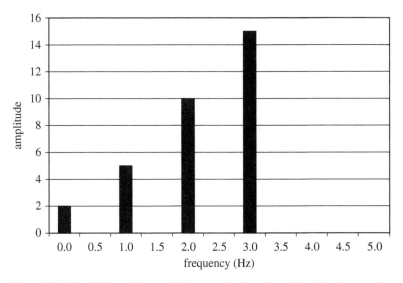

Figure 9.4   The DFT amplitudes of the example function, $x(t)$, plotted versus frequency.

9.7   Sampling and Aliasing

Consider the case where there is a single frequency harmonic signal at f Hz. That is

$$x(t) = A\cos(2\pi ft) + B\sin(2\pi ft). \tag{9.56}$$

If we used a DFT to detect this signal we would need to sample fast enough to capture the information required to calculate the coefficients a_n and b_n using Equations 9.43 and 9.45, for example. The question is, *how fast must we sample to gather the necessary information about the signal?* The answer is fairly straightforward. You can visualize the process as follows.

Since there is only a single frequency present, all but one of the terms in the summation required for the DFT of Equation 9.29 will be zero and the DFT will reduce to

$$x(t) = a_p\cos(p\omega_0 t) + b_p\sin(p\omega_0 t) \tag{9.57}$$

where $p\omega_0 = 2\pi f$. That is, only the term at the frequency of the signal will remain.

Equation 9.57 has two unknowns, a_p and b_p. Returning to our curve fitting analogy, we have two unknowns so we will need two points per cycle in order to be able to calculate the amplitudes required. We therefore need to sample at a frequency at least twice as high as the frequency to be detected.

Let Δt be the time between samples. The sampling rate is then

$$f_s = \frac{1}{\Delta t}\ \text{Hz} \tag{9.58}$$

and the highest frequency that can be detected, called the *Nyquist frequency*, is

$$f_c = \frac{f_s}{2} = \frac{1}{2\Delta t}\ \text{Hz} \tag{9.59}$$

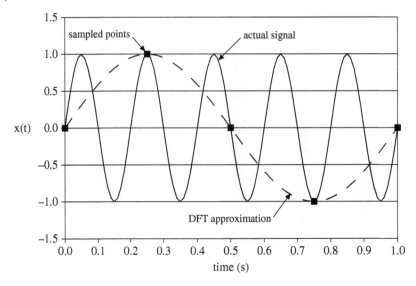

Figure 9.5 Aliasing.

The obvious next question is, *since the Nyquist frequency is the highest that we can detect with our sampling rate, what happens if the signal has frequency content above the Nyquist frequency?*

Figure 9.5 shows a high frequency signal (solid line) sampled at five points (squares) and the DFT approximation that would result (dashed line). Clearly, the high frequency signal is in the data and the sampling will detect it but at less than two points per cycle so the DFT approximation will be a signal with a lower frequency. This is called *aliasing*, a fitting name for a high frequency signal masquerading as a low frequency signal.

In order to determine what frequency we can expect to see in the DFT, consider the following and note that the complex exponential notation for harmonics is used simply because it shortens the argument. Let there be a high frequency signal, $x(t)$, with amplitude A, frequency ω, and phase angle ϕ so that we can write

$$x(t) = Ae^{i(\omega t + \phi)} = Ae^{i\omega t}e^{i\phi} \tag{9.60}$$

and let the signal detected by the DFT, $x_d(t)$, have an amplitude a, be at a detected frequency, ω_d, and have a phase angle ϕ_d

$$x_d(t) = ae^{i(\omega_d t + \phi_d)t} = ae^{i\omega_d t}e^{i\phi_d} \tag{9.61}$$

where $\omega_d < \omega$.

Let there be a time t_0 where $x(t)$ and $x_d(t)$ are equal. This corresponds to one of the sampled points shown in Figure 9.5 and can be expressed as

$$Ae^{i\omega t_0}e^{i\phi} = ae^{i\omega_d t_0}e^{i\phi_d}. \tag{9.62}$$

The two signals will be equal again at the next sampled point where $t = t_0 + \Delta t$ so we can write

$$Ae^{i\omega(t_0 + \Delta t)}e^{i\phi} = ae^{i\omega_d(t_0 + \Delta t)}e^{i\phi_d} \tag{9.63}$$

or

$$Ae^{i\omega t_0}e^{i\omega \Delta t}e^{i\phi} = ae^{i\omega_d t_0}e^{i\omega_d \Delta t}e^{i\phi_d}. \tag{9.64}$$

Equation 9.62 can be used to cancel terms in Equation 9.64, resulting in

$$e^{i\omega \Delta t} = e^{i\omega_d \Delta t}. \tag{9.65}$$

Define ω' to be the difference between ω and ω_d so that

$$\omega = \omega_d + \omega'. \tag{9.66}$$

Substituting Equation 9.66 into Equation 9.65 yields

$$e^{i(\omega_d + \omega')\Delta t} = e^{i\omega_d \Delta t} \tag{9.67}$$

or

$$e^{i\omega_d \Delta t}e^{i\omega' \Delta t} = e^{i\omega_d \Delta t} \tag{9.68}$$

from which

$$e^{i\omega' \Delta t} = 1. \tag{9.69}$$

Equation 9.69 can be written as

$$\cos \omega' \Delta t + i \sin \omega' \Delta t = 1 \tag{9.70}$$

and, by equating the real and imaginary parts on the left and right hand sides, we get two simultaneous equations, as follows

$$\cos \omega' \Delta t = 1 \text{ and } \sin \omega' \Delta t = 0. \tag{9.71}$$

The dual requirements in Equation 9.71 are satisfied by

$$\omega' \Delta t = \pm 2k\pi; k = 0, 1, 2, \ldots \tag{9.72}$$

from which

$$\omega' = \pm 2k\pi \left(\frac{1}{\Delta t}\right) ; k = 0, 1, 2, \ldots. \tag{9.73}$$

Substituting this into Equation 9.66 yields

$$\omega = \omega_d \pm 2k\pi \left(\frac{1}{\Delta t}\right) ; k = 0, 1, 2, \ldots. \tag{9.74}$$

We can change Equation 9.74 from frequencies in rad s^{-1} to frequencies in Hz by noting that $\omega = 2\pi f$ and $\omega_d = 2\pi f_d$, resulting in

$$2\pi f = 2\pi f_d \pm 2k\pi \left(\frac{1}{\Delta t}\right) ; k = 0, 1, 2, \ldots \tag{9.75}$$

Canceling the 2π terms gives

$$f = f_d \pm k \left(\frac{1}{\Delta t}\right) ; k = 0, 1, 2, \ldots. \tag{9.76}$$

We make one last change to Equation 9.76 by introducing the Nyquist frequency $f_c = 1/2\Delta t$ from Equation 9.59 to get

$$f = f_d \pm 2k f_c ; k = 0, 1, 2, \ldots. \tag{9.77}$$

Equation 9.77 specifies which high frequency components, f, will appear at the detected frequency, f_d. It is clear that, since $f_d \le f_c$, some of the high frequencies will be negative. Consider only the positive frequencies, which are

$$f = f_d + 2kf_c \; ; k = 0, 1, 2, \dots \tag{9.78}$$

and

$$f = -f_d + 2kf_c \; ; k = 0, 1, 2, \dots . \tag{9.79}$$

Equations 9.78 and 9.79 can be combined as

$$f = 2kf_c \pm f_d \; ; k = 0, 1, 2, \dots . \tag{9.80}$$

The Nyquist frequency is sometimes called the *folding frequency* because of Equation 9.80. For the case where $k = 1$, we find that the first frequency aliased with f_d is

$$f = 2f_c - f_d \tag{9.81}$$

and this frequency can be found by folding the frequency scale shown in Figure 9.6 around the Nyquist frequency, f_c.

Consider again the example of Section 9.6 but, in addition to the signal, $x(t)$, that has three frequency components, we will add another higher frequency component at 8.5 Hz. The sampling time is $\Delta t = 0.10$ s, so that the highest frequency that will be detectable is the Nyquist frequency $f_c = 1/(2\Delta t) = 5$ Hz. As a result, the 8.5 Hz signal is outside the detectable range but we expect to see it as an aliased signal at (according to Equation 9.81) $f_d = 2 \times 5 - 8.5 = 1.5$ Hz. Let the 8.5 Hz component have an amplitude of 20 so that the function in Equation 9.48 becomes

$$x(t) = 2 + 5\sin(2\pi t) + 10\sin(4\pi t) + 15\cos(6\pi t) + 20\cos(17\pi t). \tag{9.82}$$

Table 9.4 shows the coefficients that are calculated by the DFT for this revised case. Notice that a_3, the coefficient related to the cosine term at 1.5 Hz, erroneously has the amplitude of the 8.5 Hz component. Aliasing is a serious problem in experimental data analysis. Figure 9.7 shows the dramatic change from Figure 9.4.

There will invariably be signal components at frequencies higher than the frequency range of interest in the experiment. The only way to avoid aliasing is to use a low-pass analog filter when the data are recorded. The filtering has to be done using hardware. Section 9.9 discusses digital filtering after the data have been recorded but it needs to be emphasized

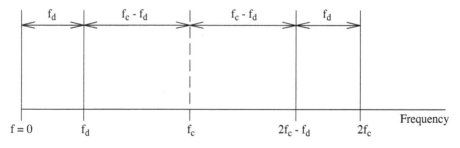

Figure 9.6 The folding frequency.

Table 9.4 DFT coefficients.

Coefficient	Value	Frequency (Hz)
a_0	2.000	0.00
a_1	0.000	0.50
a_2	0.000	1.00
a_3	**20.000**	1.50
a_4	0.000	2.00
a_5	0.000	2.50
a_6	15.000	3.00
a_7	0.000	3.50
a_8	0.000	4.00
a_9	0.000	4.50
a_{10}	0.000	5.00
b_1	0.000	0.50
b_2	5.000	1.00
b_3	0.000	1.50
b_4	10.000	2.00
b_5	0.000	2.50
b_6	0.000	3.00
b_7	0.000	3.50
b_8	0.000	4.00
b_9	0.000	4.50

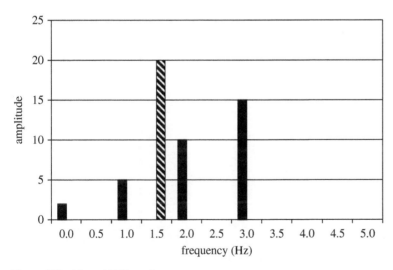

Figure 9.7 Aliased DFT results.

that digital filtering relies on accurate DFTs and, if the aliased results are in the data, the digital filtering process will not remove them because they are seen to be legitimate low frequency signals.

9.8 Leakage and Windowing

To illustrate the concept of *leakage*, we consider an example where we have a signal with two frequencies. Let

$$x(t) = 60\cos(40\pi t) + 90\cos(80\pi t). \tag{9.83}$$

That is, we have an amplitude of 60 at 20 Hz and an amplitude of 90 at 40 Hz. Choose $\Delta t = 0.01$ s and collect 100 samples. As a result we have a sample time $T = 100 \times 0.01 = 1$ s, a frequency resolution $\Delta f = 1/T = 1$ Hz, and a Nyquist frequency $f_c = 1/(2\Delta t) = 50$ Hz. The DFT is calculated and the results, shown in Figure 9.8, are as expected.

Now we consider a different signal with the same amplitudes but slightly different frequencies (amplitude of 60 at 20.5 Hz and an amplitude of 90 at 39.5 Hz)

$$x(t) = 60\cos(41\pi t) + 90\cos(79\pi t). \tag{9.84}$$

Using the same settings for the DFT as in the first signal, the result is as shown in Figure 9.9. The immediate question is, *what happened?*

Before answering the *what happened?* question, it must be pointed out that we are looking at what is called *leakage*. This is a phenomenon where the amplitudes that are present in the signal at discrete frequency bins *leak* into neighboring bins.

As to the question of *what happened?*, we can start by considering the continuous Fourier transform of a square wave. Figure 9.10 shows the square wave and some of the components

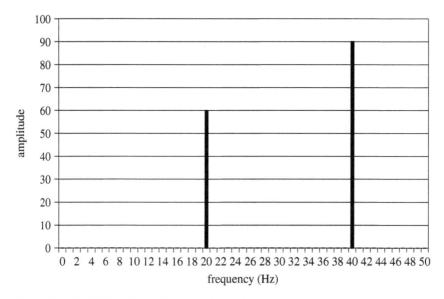

Figure 9.8 The DFT for the first example (no leakage).

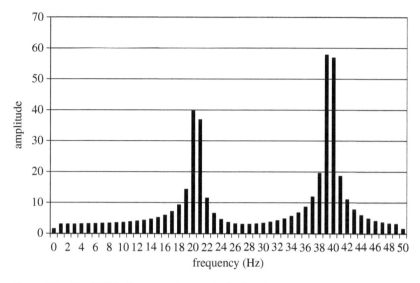

Figure 9.9 The DFT for the second example (leakage).

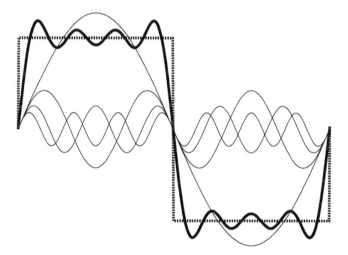

Figure 9.10 CFT approximation to the square wave.

that are used in approximating it. The thin lines are components in the expansion and the thick line is the sum of the components. Theoretically, an infinite number of components will exactly duplicate the square wave but, as can already be seen, this will require some amplitude at every frequency in order to fill in the discontinuities. The square wave has the same period as the first component of the Fourier series but the corners can only be filled in by adding higher and higher frequency components.

Going back to our two examples, the difference is that the first example (20 Hz and 40 Hz) is a signal where both components have completed an integer number of cycles in the 1 s sample that we considered. In the second example (20.5 Hz and 39.5 Hz), neither component is at the end of a cycle when the sample is truncated so the DFT sees a discontinuity and tries to deal with it by using all available frequency components.

The Fourier transform is only valid for a periodic function and requires information for complete cycles of the signal. The chances of any experimental data sample being periodic and being sampled over complete cycles is essentially zero. Leakage can't be completely eliminated but it can be reduced by *windowing* the data. Windowing forces the data to appear to be periodic by multiplying them by a function that is zero at both the start and the end and that rises to a value of one in the center.

You can imagine all sorts of windows. For example, the rectangular window goes from zero to one at the beginning of the data, stays at one, and then goes back down to zero at the end of the data. This is the default window that we get without manipulating the data so it has no effect on leakage. The simplest window that has an effect is the triangular window that ramps from zero at the start of the data to one at the center of the data and then back down to zero again at the end of the data. It is, however, more common to use windows that have continuous derivatives and zero slopes at their ends. A very common window function is the Hanning window shown in Figure 9.11

The Hanning window is expressed as a weighting function, $w(x)$

$$w(x) = \frac{1}{2}(1 + \cos \pi x) ; \ -1 \le x \le 1 \tag{9.85}$$

and is shown time-scaled in Figure 9.11 in order to cover the one second of data we have been considering in our examples.

Figures 9.12 and 9.13 show the original data and the windowed data respectively. It is clear from the change in shape of the data that the windowing will have an effect on the amplitudes of the DFT coefficients. We take the amplitude effect into account by realizing that the area under the Hanning window is one-half the area under a rectangular window. As a result, the windowed amplitudes need to be multiplied by two in order to restore their actual values. This area scaling needs to be done whichever windowing function is used.

Figure 9.14 shows the DFT for the windowed data of Figure 9.13. While not being perfect, the result is far superior to that shown in Figure 9.9.

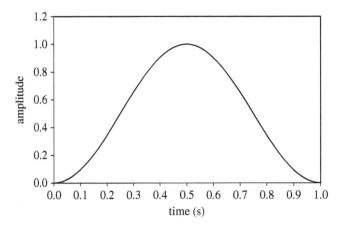

Figure 9.11 The Hanning window.

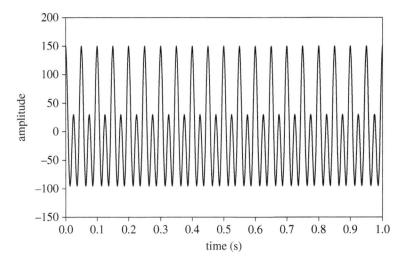

Figure 9.12 The data from Equation 9.83.

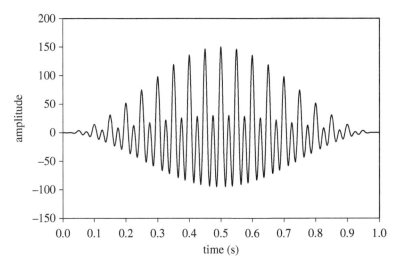

Figure 9.13 The windowed data.

9.9 Decimating Data

There are times when we realize that we have sampled at a higher rate than is necessary and we wish to reduce the sampling rate. For example (and this has happened), if we set the sampling rate in our A/D converter to 100 000 samples per second and we analyze using an FFT limited to 8192 points, we will have a total sample time $T = 0.08192$ s and the frequency resolution will then be $\Delta f = 1/T = 12.2$ Hz. If your analysis is meant to be able to distinguish between signals at 20 and 30 Hz, for instance, then you are out of luck because both of these signals will be in the same frequency bin. It becomes tempting to say that, since

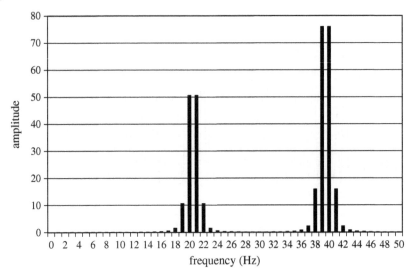

Figure 9.14 The DFT for the second example with windowing.

this is all digital data, we can simply ignore some of the data and we will get the same result as if we had sampled more slowly. That is, just using every second point will be equivalent to sampling at 50 000 samples per second or, even better, using every hundredth point will be equivalent to sampling at 1000 Hz so that 8192 points will give $\Delta f = 0.122$ Hz.

While it is true that we can leave out data points and the result will be exactly as if we had sampled at a slower rate, we need to keep in mind the possibility of aliasing as discussed in Section 9.7. If we did our work correctly when we sampled the data, we used a low-pass analog filter with a cut-off frequency we desired and then sampled at a rate that was at least twice as high as the cut-off frequency. For the example just discussed, we sampled at 100 000 samples per second so our cut-off frequency would have been around 50 000 Hz. The digital signal we recorded therefore has frequency components up to 50 000 Hz. If we then decide to use every hundredth point to set our sampling rate to 1000 Hz, we are going to get serious aliasing.

The process of leaving out data points is called *downsampling*.

There is a difference between the aliasing stemming from downsampling and that which comes from not using a low-pass filter when sampling data. If you omit the low-pass filter your DFT will interpret high frequency signals as low frequency signals and you won't be able to differentiate between actual peaks and aliased peaks. In the case of downsampling, you know where the aliased peaks actually are in the sense that you see them in the DFT before you do the downsampling. This gives you the option of using a digital filter to remove them before you do the downsampling. The sequence of using a low-pass digital filter followed by downsampling is called *decimation*.

For the example we are using, we want to get rid of anything above 25 Hz before we downsample. This is best accomplished in the frequency domain by performing a DFT on the full data set, then implementing a low-pass filter with a cut-off at 25 Hz, and then performing the inverse DFT to get a revised time signal. Implementing the filter in the frequency domain is simply a matter of using a rectangular window that is equal to one for frequencies below the cut-off frequency and zero above it. This can be accomplished by setting the DFT coefficients related to frequencies above the cut-off frequency to zero.

9.10 Averaging DFTs

The last topic to be covered in this chapter has to do with suppression of noise in experimental measurements. Noise in measurements consists of random changes in signal values that can, in some cases, reach levels that approach that of the signal you are trying to measure. It is good experimental practice to attempt to minimize noise levels before recording data. Nevertheless, there will always be some level of noise in experimental data.

Since the signal comprises two components, one of which is regular and repeatable while the other is random, it can be expected that two DFTs calculated over different parts of the time signal will give different results. Both will have large amplitudes at the frequencies of the signal but smaller and non-repeatable amplitudes at other frequencies. It can therefore be expected that if several of the DFTs are averaged, the real signal will be prominent since it appears in every DFT whereas the noise will give different peaks in different spectra and, on average, will not be prominent. This is exactly what happens. The noise is part of the signal and will never be zero but averaging the spectra results in a smoothing of the noise. This is a simple linear averaging process where the DFT amplitudes at each frequency are added and then divided by the number of DFTs performed.

Exercises

Descriptions of the systems referred to in the exercises are contained in Appendix A.

9.1 Prove that the trigonometric integrals in Equation 9.10 are zero. That is, for $\omega_0 = 2\pi/T$, show that,

$$\int_0^T \cos(n\omega_0 t)\mathrm{d}t = 0 \text{ and } \int_0^T \sin(n\omega_0 t)\mathrm{d}t = 0.$$

9.2 Show that, for $\omega_0 = 2\pi/T$ and $p \neq n$,

$$\int_0^T \cos(n\omega_0 t)\sin(p\omega_0 t)\mathrm{d}t = 0$$

$$\int_0^T \sin(n\omega_0 t)\sin(p\omega_0 t)\mathrm{d}t = 0$$

$$\int_0^T \cos(n\omega_0 t)\cos(p\omega_0 t)\mathrm{d}t = 0$$

$$\int_0^T \sin(n\omega_0 t)\cos(p\omega_0 t)\mathrm{d}t = 0$$

by using the identities

$$\cos(n\omega_0 t)\cos(p\omega_0 t) = \frac{1}{2}\cos[(n+p)\omega_0 t] + \frac{1}{2}\cos[(n-p)\omega_0 t]$$

$$\sin(n\omega_0 t)\sin(p\omega_0 t) = \frac{1}{2}\cos[(n-p)\omega_0 t] - \frac{1}{2}\cos[(n+p)\omega_0 t]$$

$$\cos(n\omega_0 t)\sin(p\omega_0 t) = \frac{1}{2}\sin[(n+p)\omega_0 t] - \frac{1}{2}\sin[(n-p)\omega_0 t]$$

$$\sin(n\omega_0 t)\cos(p\omega_0 t) = \frac{1}{2}\sin[(n+p)\omega_0 t] + \frac{1}{2}\sin[(n-p)\omega_0 t].$$

9.3 Use Equation 9.85 to show that the area under the Hanning window is one-half the area under the rectangular window.

9.4 Your data acquisition system can sample at any rate you choose between 1 and 4000 samples per second and it has sufficient memory to store up to 8192 samples. You also have a low-pass filter with an adjustable cut-off frequency that can be set anywhere between 1 and 2000 Hz. You have been asked to come into a building experiencing excessive floor vibrations and determine whether the cause is a rotary machine operating at 1800 RPM in this building or a reciprocating compressor operating at 28 strokes per second in an adjacent building.

You have an accelerometer and plan to measure and record accelerations on the floor. You will then perform an FFT and calculate the response spectrum to see which machine is at fault.

Specify your sampling rate, how many samples you should take per measurement, what your cut-off frequency should be, and how many measurements you can average. Give reasons for your choices.

9.5 Program the DFT series of Section 9.4 using software of your choice and then generate some data similar to that used as an example in Section 9.6 (i.e. Equation 9.48) and test your software on it. Experiment with the software – try out aliasing, leakage, and so on.

9.6 If you install strain gauges on the spindle of a milling machine, you will be able to pick up both the relatively constant normal force on the cutter and the cyclic forces caused by tooth engagement as the spindle turns. If you plot the total force versus time, you will see the cyclic forces as a relatively small oscillation superimposed on the large, constant, normal force. The amplitude of the cyclic force variation is about 5% of that of the normal force.
a) What is the mean value of this signal?
b) What would you expect the DFT of the signal to look like?
c) If you calculated the mean value in the time domain and subtracted it from each of the data points before doing the DFT, what would your DFT look like?
d) Using the DFT of the zero mean signal in part (c) is the preferred way of getting the frequency content of a signal in most cases. Why do you think this is?

9.7 Using the simulation data you generated for system 1 in Exercise 8.1 and your DFT series software, find the frequency spectrum. Compare the frequencies where your DFT shows peaks to the natural frequencies determined in Exercise 5.2.

9.8 Run your simulation of system 12 from Exercise 8.3 and generate enough data to be able to produce a DFT. Do you see the natural frequency?

A

Representative Dynamic Systems

A.1 System 1

Two massless rigid rods (lengths d_1 and d_2) support two particle masses (m_1 and m_2) at points A and B as shown in Figure A.1. The joints at O and A are frictionless pinned joints and the bodies move in two dimensions under the influence of gravity. Use the angles θ_1 and θ_2 as degrees of freedom. The coordinate system ($\vec{i} - \vec{j} - \vec{k}$) is fixed in link OA and $\vec{i}$ is aligned with OA.

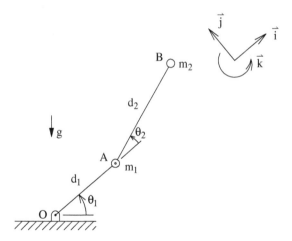

Figure A.1

A.2 System 2

A thin, uniform, rigid rod (AB) has length 2ℓ and mass M. The center of mass of the rod is at G. The rod is free to slide on the frictionless pivot at O as shown in Figure A.2. Attached to one end of the rod (B) is an inextensible string of length d that supports a particle with mass m (C). A known force $F(t)$ is applied to the other end of the rod (A) and remains

The Practice of Engineering Dynamics, First Edition. Ronald J. Anderson.
© 2020 John Wiley & Sons Ltd. Published 2020 by John Wiley & Sons Ltd.
Companion Website: www.wiley.com/go/anderson/engineeringdynamics

perpendicular to the rod at all times. The bodies move in two dimensions under the influence of gravity. The three degrees of freedom are θ, ϕ, and x.

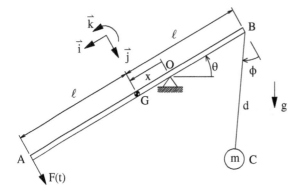

Figure A.2

A.3 System 3

A small vehicle has been designed as a Mars rover and a simple dynamic model of it is shown in Figure A.3.

The vehicle is modeled as a single rigid body supported by two vertical springs, each having stiffness, k. The lower ends of the springs are attached to small, rigid, massless wheels that exactly follow the terrain as the vehicle moves forward. The center of mass of the vehicle is at G. One spring is located a distance, a, ahead of G and the other is a distance, b, behind G. Motions of the vehicle are described by the vertical translation of the center of mass, x, and a pitch rotation, θ, with positive directions as shown. The vehicle has a total mass of m, a pitch moment of inertia about G of I_G, and moves forward with a velocity V.

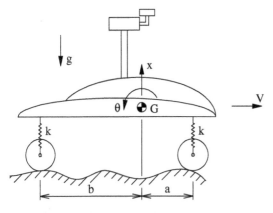

Figure A.3

A.4 System 4

The circular disk rotates in a horizontal plane about the fixed point O. It has constant angular speed Ω in the direction shown in Figure A.4. A slot has been milled into the disk in a radial direction. The small mass, m, is able to move along the slot without friction and is attached to the center of the disk by a spring of stiffness, k. The spring has an undeflected length of ℓ_0 (i.e. there is no force in the spring when it has a length of ℓ_0).

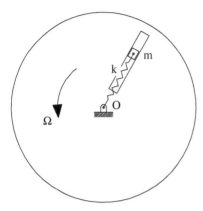

Figure A.4

A.5 System 5

Figure A.5 shows an inclined plane that is used to unload cylindrical oil drums from a platform.

The drums are rolled onto the end of the plane when it is in a horizontal position ($\theta = 0$). A person applies a vertical force, F, to the opposite end of the plane in order to pivot the plane about the frictionless pin at O, thereby raising the drums so that they roll without slipping down the plane, eventually being deposited on the ground. The force varies in magnitude as the drum rolls down the plane but is always perfectly vertical. The center of mass of the drum is at G and the center of mass of the platform is at O.

The parameters are:

Oil drum: mass $= m$, radius $= r$, moment of inertia about the center of mass $= \frac{1}{2}mr^2$
Inclined plane: mass $= m$, moment of inertia about the pivot point $= 3mr^2$.

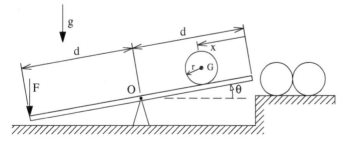

Figure A.5

A.6 System 6

Figure A.6 shows a disk with an angular velocity ω about a vertical axis. Attached to brackets at the edge of the disk is a massless rigid rod of length r that has a point mass m at its free end. The brackets welded to the disk contain bearings that allow the rigid rod to pivot freely so that the mass can move in a vertical plane as the disk spins.

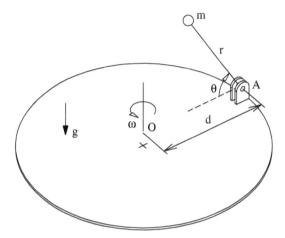

Figure A.6

A.7 System 7

The system shown in Figure A.7 consists of three rigid bodies. The two slender rods are identical, each having mass $= m$ and length $= 2L$. They are connected at their ends by frictionless pin joints at O, A, and B. The rectangular body also has mass $= m$ and slides without friction in its guideway. A constant force F is applied to this body in the direction indicated. A spring with stiffness $= k$ connects the system to ground at point A as shown. The spring is undeflected when $\theta = 0$. The system operates in the horizontal plane so that gravity has no effect on it.

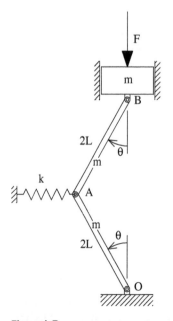

Figure A.7

A.8 System 8

Figure A.8(left) shows a rectangular object with four point masses attached to it by massless, slender, rigid rods connected to frictionless pin joints at the corners (such as at A). The object is at rest and the masses rest against it. We consider the rectangular object and the four attached bodies to constitute a *system*. The system is free to rotate in the horizontal plane about the fixed point O.

Figure A.8(right) shows the system after some applied torque has started it rotating. The rectangular object now has an angular velocity ω and an angular acceleration $\dot{\omega}$ in the direction shown and the point masses are in motion relative to it.

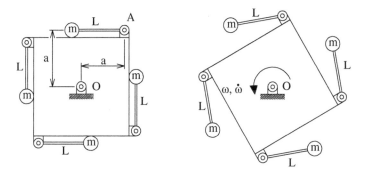

Figure A.8

A.9 System 9

Figure A.9 shows a slotted, thin, uniform rod OA that is connected to the ground by a frictionless pin at O and moves in a horizontal plane. It is also connected to the ground by a

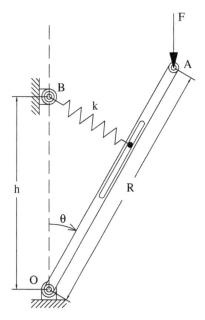

Figure A.9

linear spring of stiffness, k, that is connected to a slider in the slot. The slot is frictionless so that the spring always remains perpendicular to the rod. The spring can be modeled as having an undeflected length of zero. That is, there is no force in the spring when $\theta = 0$. The force F, applied at point A, is constant and remains parallel to line OB.

A.10 System 10

Figure A.10 shows a slender uniform rod of mass $3m$ and length d that is connected to the ground by the frictionless pin at O. The rod is connected to a linear spring of stiffness k which is in turn connected to a slider of mass m that moves in a frictionless slot. The slider is also connected to the ground by another linear spring of stiffness k. The springs are undeflected when the angle θ and the displacement x are zero. The system moves in a horizontal plane. A force $F(t)$ is applied to the free end of the rod as shown. This force is always perpendicular to the rod.

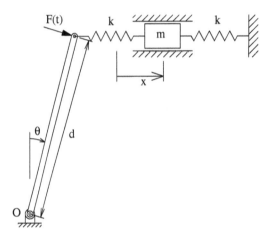

Figure A.10

A.11 System 11

Figure A.11 shows a system in two different positions. The system consists of a massless rigid rod of length L connecting two massless rollers that are constrained to move in horizontal and vertical slots as shown. Connected to the roller on the right is a mass m that moves freely in the vertical slot. The spring is undeflected when the system is in the position shown in the left-hand figure. When the system is in motion, it has a single degree of freedom θ as shown in the figure to the right. There is an equilibrium value of θ as is implied by the figure to the right.

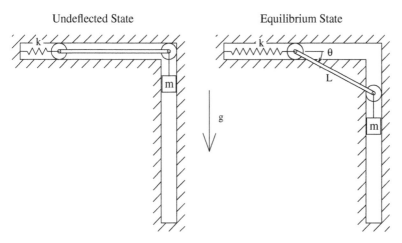

Figure A.11

A.12 System 12

Figure A.12 shows a slender uniform rod of mass m and length $2d$ that is connected to the ground by the frictionless pin at O. The rod is supported by a linear spring of stiffness k connected to it at A and to a slider in a frictionless, ground-fixed, slot at B. As the rod moves, the slider aligns itself in the slot to ensure that the spring is always truly vertical. The spring is undeflected when the angle θ is zero.

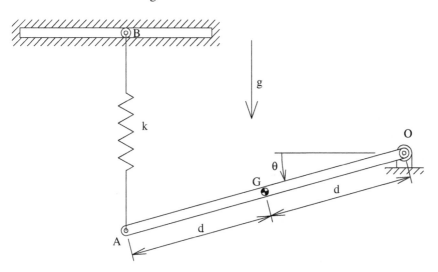

Figure A.12

A.13   System 13

Figure A.13 shows a thin uniform rod OA (mass = m, length = 2ℓ) pinned to the ground at O and to an identical thin uniform rod BC at A. Both pins are frictionless. The system moves in a vertical plane. A spring of stiffness k has one end attached to rod BC at B and the other end attached to a slider D that moves freely within a frictionless slot in OA. The slider causes the spring to remain perpendicular to rod OA at all times. The force in the spring is zero when its length is zero.

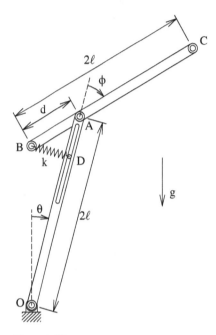

Figure A.13

A.14   System 14

Figure A.14 shows a uniform disk (mass = m; radius = r) that rolls without slipping on a stationary circular body that has radius R. The center of mass of the disk is at point A and

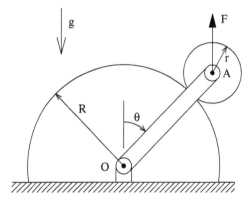

Figure A.14

the disk is pinned there to a uniform slender rod (mass $= m$) that is also pinned to the center of the stationary body at point O. The pins are frictionless. There is a constant vertical force F applied to the rod at point A.

A.15 System 15

Figure A.15 shows a particle of mass, m, that is supported in the vertical plane by two springs of stiffness, k, and, $4k$. The springs have undeflected lengths of zero. A constant horizontal force, F, is applied to the particle as shown.

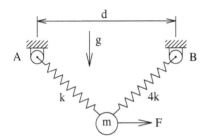

Figure A.15

A.16 System 16

Figure A.16 shows a slender uniform rod AB (mass $= m$; length $= 2\ell$) attached to a massless, frictionless roller at point A. The roller travels in the vertical plane inside a circular slot (radius $= r$) cut into a solid structure.

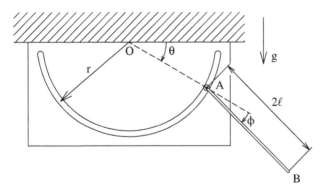

Figure A.16

A.17   System 17

The small mass, m, in Figure A.17 moves freely on the rotating circular hoop. There is *friction* between the mass and the hoop and there is *no gravity*. The hoop has a constant angular velocity vector of magnitude ω in the direction shown.

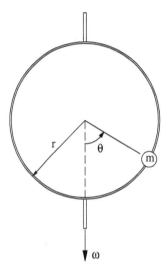

Figure A.17

A.18   System 18

Three masses are connected by three identical springs as shown in Figure A.18. The masses are constrained to have vertical motion only and the degrees of freedom, relative to the equilibrium positions of the masses, are x_1, x_2, and x_3 as shown. The vertical force, $F(t) = F \sin \omega t$, is applied as shown in the figure.

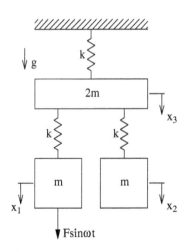

Figure A.18

A.19   System 19

Figure A.19 shows one truck of a transit vehicle with degrees of freedom specified as bounce (*x*) and pitch (*θ*) of the truck frame. Use *m* for the truck frame mass, *I* for the truck frame pitch moment of inertia, k_s for the secondary stiffness where the truck frame is connected to the carbody, and k_p for the primary stiffnesses that connect the wheels to the truck frame. The car body is massive enough that it can be assumed to have no motion and the wheels remain stationary on the rails. Assume that damping is small enough to be neglected. The dimensions *a*, *b*, and *c* are known.

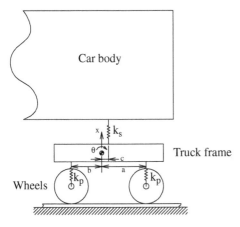

Figure A.19

A.20   System 20

The two carts (masses *m* and 4*m*) in Figure A.20 are connected together and to the ground by springs of stiffness, *k*, as shown. There is a force, *f*(*t*), applied to the lower cart.

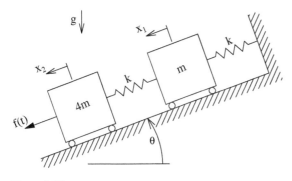

Figure A.20

A.21 System 21

Figure A.21 shows a cart of mass, m, that is attached to the ground at point A by a spring of stiffness, k, and a damper with damping coefficient, c. The motion of the cart is forced by a harmonic motion, $y(t)$, at the end of another damper with coefficient, c. $y(t)$ has amplitude, Y, and frequency, ω.

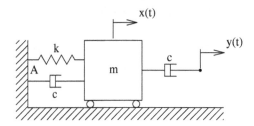

Figure A.21

A.22 System 22

Figure A.22 shows a robot moving a rectangular solid block of mass, m, with dimensions, $b \times b \times 2b$. The arm, AB, has constant length, ℓ, and rotates in a horizontal plane with angular velocity, ω_0. The arm, BC, has variable length, x, and angular velocity about the line, AB, of magnitude, ω_1.

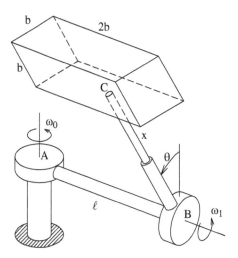

Figure A.22

A.23   System 23

Figure A.23 shows a three dimensional pendulum with a massless, axially elastic rod supporting a particle of mass, m. The system is free to move in 3D except in so far as point, O, is constrained to remain attached to the ground and the rod, Om, cannot bend. The rod has an axial stiffness, k, and an undeflected length, ℓ_0. The position of the mass is specified by the length of the rod, ℓ, the angle, θ, measured in the horizontal plane from fixed line, AB, and the angle, ϕ, measured from a vertical line in the vertical plane of the rod.

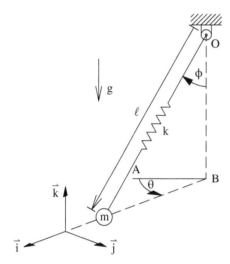

Figure A.23

B

Moments and Products of Inertia

In any problem with rotational degrees of freedom, there is a need to know the moments and products of inertia of the bodies in the system. In systems where masses are modeled as particles, the determination of the inertia properties is simply a matter of summing particle masses multiplied by products of their distances from a reference point. In rigid bodies the number of particles is infinite and the summations become integrals.

Moments and products of inertia were introduced in Section 2.9 where the definitions about a reference point O were given as,

Moment of inertia about the x-axis: $I_{xx_O} = \int_{body} (y^2 + z^2)dm$

Moment of inertia about the y-axis: $I_{yy_O} = \int_{body} (x^2 + z^2)dm$

Moment of inertia about the z-axis: $I_{zz_O} = \int_{body} (x^2 + y^2)dm$

Product of inertia about the xy-axes: $I_{xy_O} = I_{yx_O} = \int_{body} xydm$

Product of inertia about the xz-axes: $I_{xz_O} = I_{zx_O} = \int_{body} xzdm$

Product of inertia about the yz-axes: $I_{yz_O} = I_{zy_O} = \int_{body} yzdm.$

B.1 Moments of Inertia

There are several points to note about moments of inertia. First, it is clear that moments of inertia are always positive due to the fact that the integrals involve the sums of squares of distances and squared numbers are always positive. Notice that, by Pythagoras' theorem, these sums of squares are simply the square of the distance from the particle to the reference point in the plane perpendicular to the axis about which the rotation is being considered. Also notice that adding mass to a body always increases the moment of inertia unless the mass is added at the reference point.

The Practice of Engineering Dynamics, First Edition. Ronald J. Anderson.
© 2020 John Wiley & Sons Ltd. Published 2020 by John Wiley & Sons Ltd.
Companion Website: www.wiley.com/go/anderson/engineeringdynamics

It is relatively simple to show that, for a rigid body, the sum of the any two moments of inertia must be greater than the third. Simply add together I_{xx} and I_{yy} to show[1],

$$
\begin{aligned}
I_{xx} + I_{yy} &= \int_{body} (y^2 + z^2)dm + \int_{body} (x^2 + z^2)dm \\
&= \int_{body} (x^2 + y^2 + 2z^2)dm \\
&= \int_{body} (x^2 + y^2)dm + \int_{body} 2z^2 dm \\
&= I_{zz} + \int_{body} 2z^2 dm > I_{zz}.
\end{aligned}
\tag{B.1}
$$

Clearly this statement could be written for the sum of any two moments of inertia with the result that that sum must be greater than the third moment of inertia. This is a useful check on moments of inertia, especially those that are provided to the analyst by someone else. If they fail this test, then they are incorrect.

The dimensions of moments of inertia are mass times distance squared. In SI units this will be kg m^2. In US customary units, different forms, depending on the units being used for the problem at hand, are appropriate. Most commonly, one will work with lb ft sec^2 or lb in sec^2. Appendix C explains the use of different systems of units.

Because the moment of inertia is the product of mass and distance squared and the units can be somewhat complicated, it has become customary and convenient to specify the moment of inertia of a body through the use of the mass of the body and a characteristic length called the *radius of gyration*, k_O, about the reference point. The radius of gyration is defined as,

$$
k_O = \sqrt{\frac{I_O}{m}}
\tag{B.2}
$$

where m is the total mass of the rigid body and I_O is the body's moment of inertia about the reference point for the plane of motion being considered. Equation B.2 can be manipulated to read,

$$
I_O = mk_O^2
\tag{B.3}
$$

which can be interpreted as saying that the radius of gyration is the distance that a particle having the same mass as the rigid body would need to be from the reference point to have the same moment of inertia as the rigid body. The inertial properties of a rigid body are therefore often specified to an analyst as simply a mass and a distance. The actual moment of inertia is reconstructed using Equation B.3.

B.2 Parallel Axis Theorem for Moments of Inertia

The integration required to derive an expression for the moment of inertia is often tedious and to be avoided to the extent possible. Undergraduate textbooks on dynamics usually

---

1 The reference point O is dropped from the notation for brevity.

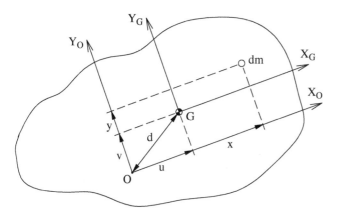

Figure B.1 Parallel axis theorem.

include extensive tables of moments of inertia for rigid bodies of various shapes. In constructing the tables, the authors have had to choose a reference point for the integration and the point chosen is most often the center of mass of the body, G. The *parallel axis theorem* allows the analyst to use the moment of inertia about the center of mass to calculate the moment of inertia about a parallel axis passing through another point without performing any integrations.

Figure B.1 shows a rigid body rotating in a plane with the center of mass, G, and another reference point, O, shown. The distance from O to G is u in the x-direction, any v in the y-direction or simply $d = \sqrt{u^2 + v^2}$.

I_{zz} can be calculated in this plane since all rotations are around the z-axis. The moment of inertia about the center of mass is,

$$I_{zz_G} = \int_{body} (x^2 + y^2)dm \tag{B.4}$$

and the moment of inertia about the reference point O is,

$$I_{zz_O} = \int_{body} [(u + x)^2 + (v + y)^2]dm. \tag{B.5}$$

Equation B.5 can be expanded and written as[2],

$$I_{zz_O} = \int [(u^2 + 2ux + x^2) + (v^2 + 2vy + y^2)]dm \tag{B.6}$$

which can be rearranged to be,

$$I_{zz_O} = \int (u^2 + v^2)dm + \int (x^2 + y^2)dm + \int (2ux + 2vy)dm \tag{B.7}$$

where we can recognize that $u^2 + v^2 = d^2$, that the second integral is the moment of inertia about G (see Equation B.4), and that u and v are constants so they can be factored out of the last integral. The result is,

$$I_{zz_O} = d^2 \int dm + I_{zz_G} + 2u \int xdm + 2v \int ydm. \tag{B.8}$$

---

2 The integrals are assumed to be over the entire body for the remainder of this discussion so the word *body* can be omitted from the integral signs to improve readability.

The first integral term in Equation B.8 is clearly just the total mass, m, of the rigid body. The two other integrals are both definitions of the location of the center of mass with respect to the origin for the integral, G, and so define the location of the center of mass with respect to itself. Both of these integrals are therefore zero and the parallel axis theorem becomes,

$$I_{zz_O} = I_{zz_G} + md^2. \tag{B.9}$$

B.3 Parallel Axis Theorem for Products of Inertia

There is also a parallel axis theorem for products of inertia. Figure B.1 shows the x–y plane of the body being considered. The products of inertia $I_{xy} = I_{yx}$ are found in this plane according to,

$$I_{xy_G} = I_{yx_G} = \int xy\,dm \tag{B.10}$$

or:

$$I_{xy_O} = I_{yx_O} = \int (u+x)(v+y)dm. \tag{B.11}$$

Equation B.11 can be expanded to give,

$$I_{xy_O} = I_{yx_O} = \int (uv + uy + vx + xy)dm \tag{B.12}$$

which, upon taking constants outside the integration, can be written as,

$$I_{xy_O} = I_{yx_O} = uv \int dm + u \int ydm + v \int xdm + \int xydm \tag{B.13}$$

For the same reasons given in Section B.2, the second and third integrals in the sequence are definitions of the location of the body center of mass with respect to itself and must therefore be zero. The first integral is the total mass of the body and the last integral is I_{xy_G}, yielding the parallel axis theorem for products of inertia,

$$I_{xy_O} = I_{yx_O} = I_{xy_G} + muv. \tag{B.14}$$

Note that the signs associated with u and v in Equation B.14 are important. The product uv can be positive or negative depending upon these signs. Referring to Figure B.1, you can see that the constant distances u and v are defined as being from the reference point O to the center of mass G along the positive X and Y axes.

B.4 Moments of Inertia for Commonly Encountered Bodies

Tables of moments of inertia are easy to find in undergraduate textbooks on dynamics. Some common moments of inertia are given here for quick reference.
Thin uniform rod: mass = m, length = ℓ

$$I_G = \frac{1}{12}m\ell^2.$$

Uniform disk: about an axis perpendicular to the plane in which it lies, mass = m, radius = r

$$I_G = \frac{1}{2}mr^2.$$

Uniform disk: about a radial axis, mass = m, radius = r

$$I_G = \frac{1}{4}mr^2.$$

Uniform rectangular block: mass = m, base = b, height = h

$$I_G = \frac{1}{12}m(b^2 + h^2).$$

C

Dimensions and Units

This appendix is included simply because of the difficulty that many people have with the concepts of dimensions and units – particularly units. *Dimensions* are measurable quantities that we use in an abstract way in that we don't specify a scale for them. They are simply referred to as length, mass, time, force, and so on. *Units* on the other hand are used to quantify dimensions using scales based on agreed upon standards. Most of the world uses SI (Système International d'unités) units while US customary units are still used in the United States.

Applied mechanics uses four dimensions – length, time, force, and mass. Three of these (length, time, and force) are measurable using common instruments such as rulers, clocks, and scales. The fourth, mass, can only be inferred from a measurement of force in the presence of a known gravitational field. When you put your tomatoes on a scale in the vegetable section of your food store and it tells you that you have 1 kg or 2.2 lb of tomatoes, the scale has actually measured the weight and not the mass. The 2.2 lb is a direct measurement of weight (i.e. force) and the 1 kg comes with the assumption that the local gravitational acceleration is 9.81 m s^{-2}, which may or may not be exactly the case since the gravitational field varies slightly around the earth. So it becomes clear that gravitational acceleration must be included in these discussions if we want to work with both mass and force. In US customary units, the gravitational acceleration is generally accepted to be 32.2 ft sec^{-2} or 386 in sec^{-2}.

A difficulty arises from the fact that the four dimensions – length, time, force, and mass – are related by a constraint equation. That is, Newton's second law, $F = ma$, must be satisfied by whatever units we choose to use. This means that only three dimensions can be independent and the fourth must be derived to satisfy Newton's second law.

The SI system specifies mass, length, and time as the independent dimensions and force is the derived unit. Mass is in kilograms (kg), length is in meters (m), and time is in seconds (s). Applying Newton's second law then requires that force be mass times acceleration where acceleration is length divided by time squared or kg m s^{-2}. This derived force quantity is called a Newton (1 N = 1 kg m s^{-2}).

US customary units are not as restrictive as SI units in that there is no specification as to what the independent dimensions are. The field of applied mechanics commonly uses force, length, and time as the independent dimensions and mass is the derived unit. Typically, but not always, force is in pounds (lb), length is in feet (ft), and time is in seconds (sec). Mass must therefore be $m = F/a$ which is lb sec^2/ft. This mass unit was once called a Slug but that term is confusing and adds a restriction to the analysis that needn't be there. It is much

The Practice of Engineering Dynamics, First Edition. Ronald J. Anderson.
© 2020 John Wiley & Sons Ltd. Published 2020 by John Wiley & Sons Ltd.
Companion Website: www.wiley.com/go/anderson/engineeringdynamics

simpler to work with units that satisfy the constraint and not give the mass unit a name at all. For example, if you are working on a vehicle dynamics problem, you are likely to measure drag force in pounds, distance in miles, and time in hours since the speed of cars in US customary units is typically measured in miles per hour (mph). The correct mass unit for this set of units is lb hour^2/mile since you simply need to generate a force in pounds when you multiply the mass by the acceleration which, in this quite ridiculous example, must be measured in miles per hour per hour or mile/hour^2. The US customary units are much more flexible than the SI units but the user needs to be careful. Rigid specifications usually lead to fewer mistakes.

A final word needs to be said about engineers who work in thermodynamics and fluid mechanics. They insist on using a quantity called the pound mass (lbm) as well as the standard pound which they label (lbf) to indicate that it is a force. The relationship between these two quantities is derived from the fact that a body weighing 1 lbf on earth ($g = 32.2$ ft/sec^2) has a mass of 1 lbm. That is, applying Newton's second law,

$$1 \text{ lbf} = 1 \text{ lbm} \times 32.2 \frac{\text{ft}}{\text{sec}^2}$$

or, dividing both sides by 1 lbf,

$$1 = 32.2 \frac{\text{lbm ft}}{\text{lbf sec}^2} = g_c$$

where g_c is a constant equal to 1 that can be used whenever it seems appropriate. My thermodynamics professor used to say "you can multiply or divide by 1 anytime without changing anything". This was advice that often caused more problems than it solved. The metric system, before it became the rigidly controlled SI system, had a kilogram force unit (kgf).

D

Least Squares Curve Fitting

This appendix is a short review of least squares curve fitting in aid of enhancing the understanding of Fourier transforms in Chapter 9.

Consider the three data points, (x_n, y_n) shown in Table D.1. We want to define a function that will best approximate these data.

There is a class of curve fitting methods called *trial functions with undetermined coefficients*. To use one of these methods, we first define a set of functions, $y_i(x); i = 1, r$, that we think are good fits to the data we have and then generate an approximation by adding together these functions multiplied by a set of coefficients, a_i, which we later determine in order to get the best possible fit to the data. That is, we let the approximation be

$$y_A(x) = \sum_{i=1}^{r} a_i y_i(x) \tag{D.1}$$

and then use one of several possible methods to find a_i. Here we restrict ourselves to using the *method of least squares*.

Consider first the case where we attempt to fit a straight line to the three points. That is, we define an approximation, $y_A(x)$, as the sum of the constant function, $y_1(x) = 1$, and the linear function, $y_2(x) = x$, multiplied by undetermined coefficients, a_1 and a_2, yielding

$$y_A(x) = a_1 + a_2 x. \tag{D.2}$$

The goal is to find the "best" possible values of a_1 and a_2. The "best" values are those that minimize the error at the three data points where we define error as the difference between $y_A(x_n)$ and y_n. The error at the nth data point is then

$$e_n = (a_1 + a_2 x_n) - y_n. \tag{D.3}$$

We can't minimize the errors as shown in Equation D.3 since they have no lower limit. Negative errors, no matter how large, would be deemed to be superior to small positive or negative errors in any minimization attempt.

The Practice of Engineering Dynamics, First Edition. Ronald J. Anderson.
© 2020 John Wiley & Sons Ltd. Published 2020 by John Wiley & Sons Ltd.
Companion Website: www.wiley.com/go/anderson/engineeringdynamics

The solution to this problem is to define a function of the errors that is always positive and then minimize that. The least squares method uses the squared errors as the function to minimize. We therefore define a function, J, as

$$J = \frac{1}{2} \sum_{n=1}^{N} (e_n)^2 = \frac{1}{2} \sum_{n=1}^{N} (a_1 + a_2 x_n - y_n)^2 \qquad \text{(D.4)}$$

where N is the number of data points.

To minimize J, we simply take the partial derivatives of J with respect to the undetermined coefficients, in this case a_1 and a_2, and ensure that they are both zero. If we think of a three dimensional plot with J plotted vertically against a_1 and a_2 axes, it would appear as a bowl with the two partial derivatives being simultaneously zero at the bottom of the bowl where J is a minimum.

The partial derivatives are

$$\frac{\partial J}{\partial a_1} = \sum_{n=1}^{3} (a_1 + a_2 x_n - y_n)(1)$$

$$= a_1 \left[\sum_{n=1}^{3} (1) \right] + a_2 \left[\sum_{n=1}^{3} (x_n) \right] - \left[\sum_{n=1}^{3} (y_n) \right] \qquad \text{(D.5)}$$

and

$$\frac{\partial J}{\partial a_2} = \sum_{n=1}^{3} (a_1 + a_2 x_n - y_n)(x_n)$$

$$= a_1 \left[\sum_{n=1}^{3} (x_n) \right] + a_2 \left[\sum_{n=1}^{3} (x_n^2) \right] - \left[\sum_{n=1}^{3} (x_n y_n) \right]. \qquad \text{(D.6)}$$

Substituting the values for x_n and y_n from Table D.1 results in the set of linear algebraic equations

$$\begin{bmatrix} 3 & 11 \\ 11 & 59 \end{bmatrix} \begin{Bmatrix} a_1 \\ a_2 \end{Bmatrix} = \begin{Bmatrix} 163 \\ 963 \end{Bmatrix} \qquad \text{(D.7)}$$

with the solution yielding

$$\begin{Bmatrix} a_1 \\ a_2 \end{Bmatrix} = \begin{Bmatrix} -17.42 \\ 19.57 \end{Bmatrix} \qquad \text{(D.8)}$$

so that the least squares linear approximation is

$$y_A(x) = -17.42 + 19.57x \qquad \text{(D.9)}$$

which is the straight line plotted on Figure D.1 where the three dots are the points from Table D.1.

Table D.1 Sample data points.

n	x_n	y_n
1	1	9
2	3	31
3	7	123

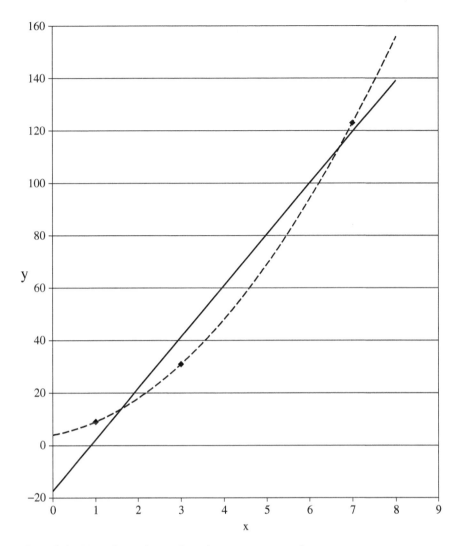

Figure D.1   Three data points and two least squares curve fits.

Also plotted on Figure D.1 is a dashed line that goes through all three data points exactly. This is the result of using the least squares method with three undetermined coefficients. That is, let the approximation be

$$y_A(x) = a_1 + a_2 x + a_3 x^2.$$ (D.10)

Formulating J and setting the three partial derivatives to zero give the set of equations

$$\begin{bmatrix} 3 & 11 & 59 \\ 11 & 59 & 371 \\ 59 & 371 & 2483 \end{bmatrix} \begin{Bmatrix} a_1 \\ a_2 \\ a_3 \end{Bmatrix} = \begin{Bmatrix} 163 \\ 963 \\ 6315 \end{Bmatrix}.$$ (D.11)

Solving Equation D.11 yields

$$y_A(x) = 4 + 3x + 2x^2 \tag{D.12}$$

which is the function used to generate the data points in the beginning.

The fact that the method of least squares will fit a curve that goes through all the data points if the number of undetermined coefficients is equal to the number of data points is relied upon in Chapter 9 where the trial functions are sines and cosines and the result is the discrete Fourier transform.

Index

The Practice of Engineering Dynamics, First Edition. Ronald J. Anderson.
© 2020 John Wiley & Sons Ltd. Published 2020 by John Wiley & Sons Ltd.
Companion Website: www.wiley.com/go/anderson/engineeringdynamics